This book is to be returned on or before the last date below.  54442   £30·00

# COMPUTERS IN CONSTRUCTION PLANNING AND CONTROL

# TITLES OF RELATED INTEREST

*Boundary element methods in solid mechanics*
S. L. Crouch & A. M. Starfield

*The boundary integral equation method for porous media flow*
J. A. Liggett & P. L-F. Liu

*Data acquisition for signal analysis*
K. G. Beauchamp & C. Yuen

*Digital methods for signal analysis*
K. G. Beauchamp & C. Yuen

*Numerical methods in engineering and science*
G. de Vahl Davis

*Plastic design*
P. Zeman & H. M. Irvine

*Structural dynamics*
H. M. Irvine

*Theory of vibration with applications\**
W. Thomson

\* Available in North America from Prentice-Hall Inc.

# COMPUTERS IN CONSTRUCTION PLANNING AND CONTROL

Michael J. Jackson

London
ALLEN & UNWIN
Boston        Sydney

© Michael J. Jackson, 1986
This book is copyright under the Berne Convention. No reproduction without permission. All rights reserved.

**Allen & Unwin (Publishers) Ltd,**
**40 Museum Street, London WC1A 1LU, UK**

Allen & Unwin (Publishers) Ltd,
Park Lane, Hemel Hempstead, Herts HP2 4TE, UK

Allen & Unwin Inc.,
8 Winchester Place, Winchester, Mass. 01890, USA

Allen & Unwin (Australia) Ltd,
8 Napier Street, North Sydney, NSW 2060, Australia

First published in 1986

**British Library Cataloguing in Publication Data**

Jackson, Michael J.
  Computers in construction planning and control.
1. Construction industry – Management – Data processing   2. Industrial project management – Data processing   3. Microcomputers
I. Title
624'.068'4         HD9715.A2
ISBN 0-04-624010-1

**Library of Congress Cataloging in Publication Data**

Jackson, Michael J. (Michael James)
  Computers in construction planning and control.
Bibliography; p.
Includes index.
1. Construction industry – Management – Data processing. I. Title.
TH438.J27      1986      624'.068      85-30843
ISBN 0-04-624010-1 (alk. paper)

Set in 10 on 12 point Times by Paston Press, Norwich
and printed in Great Britain by Butler and Tanner Ltd, Frome and London

*For Janet*

# PREFACE

This book is an exposition of a line of thought which has developed over the past seven years, the period during which I have worked in the Department of Civil Engineering at the University of Newcastle-upon-Tyne, but its roots lie deeper. All of us who have worked in the construction industry are conscious of the importance of most of the work we do, and of its expense. We are also conscious of the inefficiency of much of our work and of the cost of that inefficiency in lost expenditure, lost opportunity and lost tempers. Those of us who have moved from the industrial to the academic sphere hope to be able to reduce these losses through better techniques and better training for managers of the future.

This motivation to bring improvement has combined, in my case, with a leaning towards computer applications which has also developed throughout my career, but especially during the period I spent in the early 1970s working with the Merrison Committee on the formulation of the design rules for steel box girder bridges, an exercise almost perfectly designed to show the strengths and weaknesses of computer aid to designers.

The book is being written for a specific purpose: to enable the work that I have done at the University to be continued within the industry. The construction industry is a relatively secretive place and the knowledge gained in one company, particularly in the emotive fields of planning, control, and estimating, is rarely quickly disseminated. I hope that this book may at the least contribute to a wider discussion of techniques within companies.

Finally, I must offer my thanks to my colleagues for their encouragement, to Mrs Diane Baty for her superhuman patience in typing and retyping parts of the manuscript, and, most importantly, to my family for enduring a father who has been somewhat unendurable during these last months.

Michael Jackson

# CONTENTS

Preface — *page* ix

## 1 Planning and control in the construction industry — 1
- Planning and control — 1
- Planning and control – the differences — 3
- The special needs of project management — 5
- The special needs of construction management — 6
- The way ahead – an appropriate time for change — 7

## 2 The design of project plans — 9
- The design cycle – similarities between structural design and project planning — 9
- Pointers to change — 14
- Analytical design – a hierarchy of detail — 14
- Design detail — 16
- Planning as design – the requirements of a system — 16

## 3 The network model — 17
- Models — 17
- Network models — 17
- The conventional network representations — 19
- The analysis of simple networks — 23
- Criticality and float — 25
- The network model — 25

## 4 Graphics as an aid to interaction — 26
- The need for graphics — 26
- Graphical output — 28
- Graphical input — 33
- Graphical program direction — 36
- The use of graphics — 37

## 5 Some non-graphical planning aids — 38
- Improvements to realism — 38
- Complex links — 38
- Project closedown — 40
- Window times — 41
- Calendar dates — 41

| | | |
|---|---|---|
| | Seasonal variations in production | 42 |
| | The contribution of graphics to these techniques | 43 |
| | The pursuit of reality | 44 |

## 6  Network analysis on the computer — 45

- The sort algorithm — 45
- The forward pass for simple networks — 49
- Incorporation of window times — 51
- Complex links — 52
- Closedown periods and seasonal production variations — 57
- The part of analysis in computer-aided planning — 57

## 7  A planning program with graphical I/O — 58

- The objectives — 58
- Data input routines — 62
- Graphical output — 70
- Conclusions to be drawn — 70

## 8  Programming the graphics — 74

- Menu direction — 74
- Graphical input — 76
- Graphical output — 81
- The programming of graphics — 84

## 9  The uncertainty of construction planning — 85

- The sources of uncertainty — 85
- Increasing the accuracy of data — 85
- Inherent uncertainties — 86
- Handling uncertainty – present practice — 87
- Handling uncertainty in planning – alternative approaches — 88
- Levels of uncertainty — 89
- The use of uncertainty — 93
- The use of uncertainty in practice — 95

## 10  The analysis of uncertainty — 96

- The two approaches — 96
- The statistical method — 97
- Interfering critical paths — 98
- Simulation techniques — 101
- Data storage within simulation analyses — 102
- The speed of simulation analyses — 105
- Simulation – a viable design tool — 106

## CONTENTS

### 11 The simulation of uncertainty — 107
- The viability of simulation — 107
- The difficulties of simulation — 108
- A simulation program — 110
- The simulation program in operation — 114

### 12 Hierarchical structures and their application to project management — 116
- Projects and complexity — 116
- Plans as hierarchies — 117
- The use of hierarchies to handle complexity — 118
- The hierarchical nature of design — 118
- The hierarchy of organisation — 121
- The flow of information within a hierarchy — 123
- The application of hierarchical concepts within computer systems — 124

### 13 The hierarchical division of projects — 125
- Activity grouping — 125
- Divisional grouping — 125
- Functional grouping — 126
- Matrix grouping — 127
- The occurrence of cross links in hierarchical plans — 128
- The nature of logic and resource links — 130
- The building of hierarchies — 131
- A practical approach to the specification of hierarchies — 131

### 14 The storage of data for hierarchical planning — 133
- The computer representation and analysis of a hierarchical structure — 135
- Index lists — 138

### 15 A hierarchical program using interactive graphics — 140
- The use of graphics — 140
- The specification of resources — 142
- The analysis of hierarchical networks — 144
- The ordering of elements within hierarchical networks — 146
- The specification of cross links — 148
- Hierarchical networks – potential — 149

### 16 The control of projects — 150
- The control process — 150
- Feedforward control systems — 151

## CONTENTS

| | |
|---|---|
| Short-term forecasting | 152 |
| Construction control systems at present employed | 153 |
| The use of short-term forecasting in the industry | 155 |
| The need for computers | 155 |

### 17 The introduction of computer-aided control — 157

| | |
|---|---|
| The computer environment | 157 |
| The requirements of a construction control system | 158 |
| Site data sources | 159 |
| Data capture in the computer environment | 160 |
| An interim approach | 162 |
| Providing the forward view | 163 |
| The computer-based manual reporting system | 164 |
| Providing flexibility | 165 |
| The future | 165 |

### 18 The need for and achievement of cost estimating accuracy — 167

| | |
|---|---|
| Competitive tendering and company policy | 167 |
| The implications of low bid acceptance | 169 |
| Company policy | 171 |
| Achieving accuracy in analytical estimates | 172 |
| Bill itemisation | 173 |
| The make-up and accuracy of large value items | 175 |

### 19 Computer-aided estimating — 177

| | |
|---|---|
| Cost modelling | 177 |
| The achievement of accuracy | 178 |
| Methods of estimating | 178 |
| The role of the computer in cost estimation | 180 |
| An interactive estimating system | 181 |
| The allocation of unit rates | 181 |
| The specification of resources | 182 |
| Computer-aided estimating | 183 |

| | |
|---|---|
| Appendix   A glossary of graphics instructions | 184 |
| Further reading | 185 |
| Index | 187 |

# 1
# PLANNING AND CONTROL IN THE CONSTRUCTION INDUSTRY

**Planning and control**

In 1947 Norbert Wiener was looking for a name for the emerging science of control and decided on 'cybernetics', a word derived from the Greek word meaning steersman. In choosing this term he suggested a very useful picture of the control of projects, that of the project as a journey between two points. Figure 1.1 illustrates this analogy – an analogy which can shed considerable light on the function of the project manager and the systems which serve him.

As Figure 1.1 shows, the project is considered as a voyage between two points (X and Y), the method of work is the route to be taken, the route map is the project plan, and the manager is the navigator who will steer the ship. The

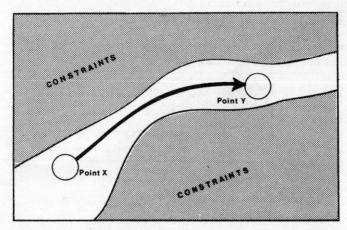

**Figure 1.1**  The journey analogy.

## PLANNING AND CONTROL IN THE CONSTRUCTION INDUSTRY

journey to be undertaken is complicated by hazards, which are analogous to the constraints imposed on the manager, and the role of the manager is to complete the journey using the resources available to him. The manager's success will depend upon the completion of the journey and its cost.

The analogy helps to identify and emphasise several features of project management which will recur in this book. The most important of these include:

**(a)** *The interdependence of planning and control.* The navigator who wishes to land safely in the intended harbour must constantly check his position against his charts, deviations from his expected route must be noted and corrective action must be taken. Steering does not occur until the direction of the ship is altered. The manager must similarly know his current position and must, if necessary, make adjustments to his plan. Control is thus impossible without a plan against which a check can be made, and planning can be seen to be a function which extends into the control phase. Effective control exists only if corrective action is taken and if this action changes things.

**(b)** *The required detail of plans.* The navigator of a ship needs very little information concerning what is astern, and little concerning what lies a great distance ahead. He needs information in great detail concerning the next mile or so. Similarly the manager will not be concerned with the detail of events and decisions which are in the distant future; he will give attention to the strategy of the future, but he will need detail for the detailed decisions which lie immediately ahead of him. The manager will only need information about events in the past to the extent that it provides a clue to what may happen in the future. This variation in the amount of detail needed by the project manager, represented in Figure 1.2, has the distinctive form of a wedge, the information wedge moving forward through time as the project proceeds. In order to carry out the project the manager needs a considerable volume of data, but he needs it in appropriate quantity at appropriate times. The generation of enormous

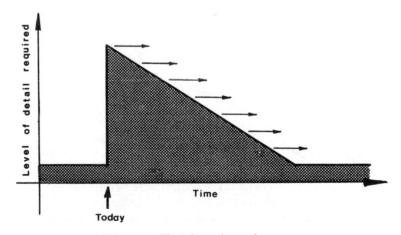

**Figure 1.2** The information wedge.

volumes of information prior to the start of the project is not helpful to the manager and is probably not cost effective, for plans change and render detail obsolete.

(c) *The need for forward vision.* Navigators get the majority of their information from looking ahead; very little is gained by observing where the vessel has been. The primary purpose of a management control system is to provide the manager with information to which he can effectively respond, and as the decisions he must make concern the future, it is information concerning the future which he needs to support his decisions. The classical approach to project control has relied on the concept of feedback, familiar in the fields of electronics and mechanical engineering, but, as the analogy shows, this concept may not be appropriate for management systems and it may be preferable to move to what Koontz and O'Donnell call a feedforward system.

(d) *The need for rapid reporting.* The major reason for doubts about the efficacy of feedback systems in the field of project management is the time taken by the control cycle. Data must be observed, compared, and acted upon, and if this process is long the information will be received too late for corrective action to be taken. By analogy the lookout at the masthead must react quickly if the ship is to avoid the iceberg; if the reporting time is very long, then either the lookout is wasting his time or he must be asked to look further ahead (with an accompanying loss of accuracy). Large and complex organisations contain long information paths and these produce slow response. Slow response jeopardises any control system even if it is future based.

(e) *The dynamic nature of plans.* A navigator must have the ability to steer his vessel. Even the most familiar voyage cannot be preset into the ship's steering mechanism. Similarly it is only in the most exceptional circumstances that a planner will be able accurately to predict all the constraints under which the project will be carried out. He will seldom find it economic or helpful to provide all the detail required by management. Plans must therefore be changed and expanded during the currency of the project. If plans are not seen to be dynamic in this way they risk being obsolete from the moment they are produced.

## Planning and control – the differences

There are important similarities between the planning and the control functions in project management and both are similar to the familiar cyclical process of design. Figure 1.3 shows the classical design cycle compared with the cycle of feedback control which is often used as a model for project management.

It is clear that, taking advantage of the slightly ambiguous nomenclature of the figure, the illustrated design cycle could refer either to the design of permanent works or products or to the design of a working method (i.e. project planning). The similarity between planning and other design activities can be further demonstrated by the consideration of some of the well-known defi-

# PLANNING AND CONTROL IN THE CONSTRUCTION INDUSTRY

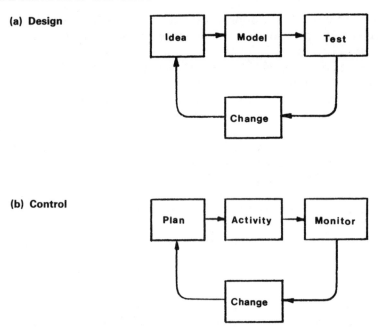

**Figure 1.3** The design and control processes; (a) design; and (b) control.

nitions of the design process. One must suffice here. Asimow defines design as 'decision making in the face of uncertainty with high penalties for error', a definition which obviously includes the risky operation known as project planning. This similarity will be pursued in the next chapter.

Consideration of Figure 1.3 suggests a strong similarity between design (=planning) and control – again Asimow's definition could be said to apply. This similarity can best be shown by removing the barrier which often separates the management functions planning and control, making the latter a continuation of the former.

The removal of this barrier is often difficult for the construction industry, where different groups of people may be responsible for the two functions, but it presents no theoretical problem. A project plan is designed using the cycle illustrated in Figure 1.3 using the best data currently to hand. Some of the data will be very reliable, some will not, and the planner must accept this variability. As the project progresses beyond the planning stage and enters the stage of actual production, the design continues; it is not frozen at the point when the action starts. Data from the site will flow back to the method designer (who is now the controller) who in the light of the improved data revises his plans.

The mechanism here is precisely the same as that whereby the planner, prior to construction, obtains additional data through sinking more boreholes or carrying out method studies. The difference is that the control function exists in time, the gathering of information and the making of decisions concerning it are

## PROJECT MANAGEMENT SPECIAL NEEDS

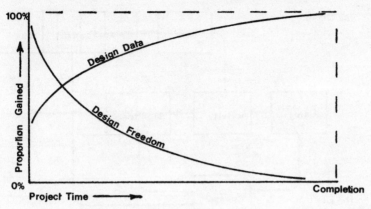

Figure 1.4  The flow of design data.

themselves activities in the network of activities which make up the project. Pre-project planning exists outside project time.

The time dimension of the control function is thus the difference between two otherwise identical processes. The same tools should be used for the two functions, and they could helpfully be regarded as one continuous project method design process in which the data available to the designer increase and his freedom of action decreases as the project proceeds from the start of the planning to project completion. The flow is illustrated by Figure 1.4.

## The special needs of project management

The difference between project management and the management of other activities is the element of uniqueness. Projects may or may not be a part of a series of similar exercises, but they will never be identical with those which have happened in the past or with those which are yet to be carried out. This difference has important effects on both the planning and the control phases of method design.

At the planning stage the familiar technique of network analysis will probably be used. In the modelling of continuous processes by networks it is usual and convenient to represent the various elements of work, be they car bodies, plastic ashtrays, or ship sections, as moving through the network. The activities are stable in time and the work flows across them. In project planning this is not the case.

In project management each activity is usually carried out once and once only. In this case it is not pieces of work which flow across the network but time. Thus all the management parameters associated with the project – the resources of labour, plant, and materials, and financial requirements, for example – are constantly varying. The maintenance of a steady resource level, virtually

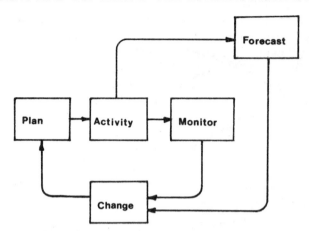

**Figure 1.5** Control through short-term forecasts.

automatic in the case of a continuous process, may be a prime and difficult objective for the project manager.

It is, however, in the control phase of a project that the difficulties become most acute and obvious. If, during the control of a continuous process, problems arise in quality or production, then, subject to the usual statistical checks and balances, the process can be altered so that the required standards of production are regained. In this environment a feedback control system will work, albeit with some wastage, and a formal feedforward system, perhaps monitoring input quality, will be very effective. In the sphere of project management, control systems, because of the one-off nature of the enterprise, must be markedly different.

If the work is not to be repeated, as is almost always the case in project management, it must be correct first time – there is no opportunity for a learning process to take place. In these circumstances, as has been said above, feedback is effective only if it is almost instantaneous, and this is rare indeed in large organisations. Adequate control can only be exercised through some form of feedforward, but even here difficulties arise because of the varying importance of the different input parameters.

One possible way of providing a form of feedforward in the project environment is to look forward to output rather than backwards to input. Such systems have been tried and have been found to be useful for alerting higher management to anticipated problems. Figure 1.5 shows how such a system compares with a normal feedback arrangement.

## The special needs of construction management

The construction industry shares with general project management the problem of the uniqueness of its activity, but it must additionally face difficulties which are special to itself. These special difficulties arise from the construction environment. Construction never takes place *in situ* and very rarely, therefore, is it possible to maintain a constant workload for a construction team in a particular location. Thus the project planner must constantly face the problem of forecasting performance in unfamiliar conditions.

The planner of construction activity is faced with more uncertainty than is his colleague in general project management. He often works with a labour force of unknown skill, a management team which has been brought together for one project only, material suppliers of doubtful reliability, and clients' representatives of unknowable rigidity. In addition to these sources of uncertainty, all of which emanate from the human resources available, are the uncertainties due to the physical environment itself. Much of construction is concerned with ground conditions, which are notoriously difficult to forecast and often very variable, and with the weather, which is impossible to forecast far in advance and can so affect site conditions as to make the most carefully made plans useless. These are only two, although admittedly the most important, of the variables due to the location which complicate the planning function. Increasing uncertainty as they do, these factors place very real limits on the cost effectiveness of highly detailed pre-project planning and on the possible accuracy of pre-project estimates.

The environment, both human and physical, affects not only planning but also control. As has been pointed out, the construction industry is obliged to work with a highly transient workforce. This affects the management of construction projects in several ways, the motivation of labour, its identity with the enterprise and its training being some of the more obvious. The transience of the labour force and its relative lack of sophistication adds to the difficulties inherent in project management of establishing and maintaining an effective control system for a project which may be very extensive, for in these conditions management may have to rely on first level supervisors or the production workers themselves to provide control data.

The physical size of the construction project affects not only the quality of the data but also the speed with which it is gathered and distributed. Information paths become long and the response time of the system increases. Control systems which, although not ideal in other fields of project management, do continue to operate, may well fail due to these extra demands.

The project management of the construction industry presents special problems for both the planner and the controller. Any attempt to discuss planning and control systems or to design new approaches to them must recognise these problems and attempt to solve them.

## The way ahead – an appropriate time for change

Many of the problems described above remain as problems not because the theoretical basis for their solution has not existed but because the necessary tools have not been easily accessible. This is not always so, however, some problems being an inherent part of construction activity, and in these cases the enthusiast for innovation should think carefully before he makes grandiose claims for the new technology. Nevertheless new resources are becoming available for the designer, and the planner of construction should be prepared to exploit these new resources in the same way that his colleagues in other engineering design disciplines are exploiting them.

The rapid drop in the cost of computing power has resulted in a parallel increase in the scope and availability of computer facilities within the construction industry. This increase suggests that this may be an appropriate time for the reappraisal of the application of computers to project planning and control. Discussions within the industry and the frequent publicity of quite minor applications show the widespread interest in such a reappraisal, although as yet little of its fruit has been made available through publication. Much of the effort has been 'in house' and designed to meet in-house needs, and such work is not widely publicised in an industry as competitive as construction. Other of the work has been directed towards the rewriting of the techniques of the 1970s so as to fit them on the small, cheap machines of the 1980s. This rewriting is certainly worthwhile, for the microcomputers which became available around 1980, being both cheap and robust, brought the possibility of site-based computer-aided planning and control, and thus contributed in a major way to breaching the walls which separate the user from the machine and the planner from the controller.

As the power and complexity of the machines available to the manager increase, so do his expectations. The development of planning and control systems categorised by the miniaturisation of the large systems of the last decade is virtually complete. The programs that are available have facilities which even five years ago could be found only in complex systems designed for mainframe machines, but hardware is moving on and even these excellent programs are not now using microcomputer hardware to the full or meeting the expectations of managers used to seeing high-quality graphical displays on their domestic television screens.

It is in the field of graphics that the major hardware advances are being made at the lower end of the computer market – the end of interest to those wishing to instal computers onto construction sites. These advances have brought high-resolution graphics from the research laboratory to the point where they are breaking into the home computer market. This movement has, of course, both been encouraged by and itself has encouraged the growth in importance of computer-aided design techniques in several divergent engineering disciplines. It is this quantum leap which makes the reappraisal of computer use particularly appropriate at this time.

# 2
# THE DESIGN OF PROJECT PLANS

It has been argued in the previous chapter that the planning and control of a project is similar in many respects to the design function in other engineering disciplines. This similarity will be used in this chapter to help to identify the features which should be offered to the planner by a useful computer-aided design (CAD) system.

**The design cycle – similarities between structural design and project planning**

The design process is familiar and has been well described in an extensive literature on the subject. It can best be represented as a cycle around which the designer proceeds, each circuit bringing him nearer to a satisfactory solution. It will later be shown that an accurate representation of the design process can be achieved only by adding a third dimension to this cycle, but present purposes are served by remaining with the traditional diagram (Fig. 2.1), which has only two dimensions.

As the diagram shows, the design process can be said to consist of three tasks,

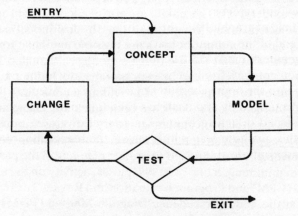

**Figure 2.1** The classical design cycle.

9

model, test, and modify, each carried out in sequence and each carried out many times as the designer approaches his objective by a process of iteration. It is instructive to examine these tasks as they are used by the project planner and his colleagues in the more usual design disciplines, for example in structural engineering.

The first process in the cycle is the building of a model representing the ideas of the designer. In the past the structural engineer used physical models in order to test his ideas. Models were laboriously built and were tested using weights. If modifications were required these were done, again laboriously, by altering the model. In some cases it can be said that full-scale modelling was being done as, for example in medieval church building, the designers erected new designs and learnt from their sometimes catastrophic mistakes. Scale models were clearly a more satisfactory design tool and these were, and still are, used in the design of the more complex structures. Even in the case of scale models, however, the cost of the design process is very high and the time required for modification makes the use of many iterations impractical.

Iteration is at the heart of the design process, and anything which makes iteration more difficult makes design less creative. Physical model building except in the most simplistic and idealised circumstances is a very slow process and real iteration cannot take place using it.

The solution to the problem of model cost and the time required for model modification comes with the replacement of physical models by mathematical models. For the structural engineer the structure can be represented by a series of equations which, when solved, give a picture of one aspect of the behaviour of the structure. Clearly such models must be idealised, for it would be difficult and, for most purposes, unnecessary to represent each detail of the completed structure, the grains of aggregate in the concrete or the grain of the timber. The designer must decide at the time when he is building his model what his objective is and must build his model with this objective firmly in mind. A large part of the skill of the designer lies in the choice of an appropriate model, the setting of a useful (and cost effective) degree of detail, and the interpretation of the results the model gives. These skills are also needed by the project planner.

For the project planner the option of the physical model has never been realistic. 'Dry runs' of important tasks are possible and have sometimes been carried out in the past, but the use of scale models, although it may appeal to the small boy in the planner and be very therapeutic, has not been practised. Again the solution to the problem of providing models finds its solution in mathematics. Mathematical models are both quick and cheap and enable the designer, in this case the project planner, to try various solutions to his design problem and so approach an optimum answer. The modelling technique which has almost universally been applied to project planning in the past two decades is network programming in one of its two guises, activity on the arrow (Critical Path Method (CPM) and Program Evaluation and Review Technique (PERT)) and activity at the node (Precedence Diagram Method (PDM)). These rival systems are shown in Figure 2.2.

## THE DESIGN CYCLE

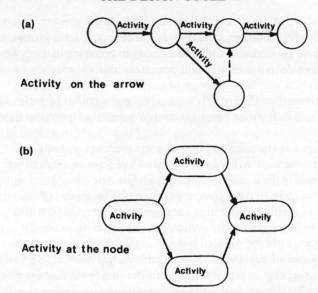

**Figure 2.2** The rival network models; (a) activity on the arrow; and (b) activity at the node.

CPM and PERT were the first network techniques to become available when they were introduced in the late 1950s. They have developed a 'brand loyalty' in the construction industry (many refer to network techniques of all types as PERT networks). Precedence networks, developed almost ten years after the others, have advantages over them which have been described elsewhere but have not yet been widely adopted because of the ten-year lead of their rivals.

Networks provide a mathematical modelling system which is analogous to those used in other design fields and the same difficulties apply. The user of networks must identify the objective of his work and should not try to produce an 'omnicompetent model', he must know the level of detail which is appropriate for his particular task, and he must know how to interpret the results which the model produces. These requirements have not always been met by the users of networks, and networking techniques have sometimes been harshly judged by the industry because they were badly used.

The second stage of the design process is the testing of the model.

The testing of the design implies a model, as has been discussed, and a criterion of success, and the setting of success criteria has led to considerable discussion within design communities. Discussion has concerned both the difficult interplay of conflicting criteria and also the definition of success levels within a single criterion.

All designers are faced with conflicting objectives and the designer for whom one criterion is so overwhelmingly important that it swamps all others is indeed fortunate (the design of long span bridges is perhaps one such discipline). For normal structural design safety, cost, appearance, durability, and maintenance cost are some of the requirements which the designer must satisfy simultaneously not only for the whole structure but also for each of the structural elements. He will do this by selecting those criteria which are dominant and will carry out his design initially with regard only to those while using his experience to avoid design decisions which will adversely affect the other, as yet untested, criteria. When a suitable stage is reached he will check his idea, using the process illustrated by Figure 2.1, with reference to other criteria. Thus the real design task is much more complex than a simple progression around a fixed cycle.

In the same way the project planner must meet several objectives simultaneously. Time and cost will often be considered to be paramount, but other criteria, plant idle time, storage capacity on site, the stability of labour demand, may also from time to time govern his design. The model of the project which the planner uses must allow him easy access to information about his performance with reference to these other, secondary, criteria, and it is likely that a model designed only for time/cost optimisation will not allow this.

The definition of success is more controversial than is the specification of criteria. In structural engineering, where the major success criterion is almost always that of safety against collapse, two opposing systems have been used. The traditional system is that which sets a limit to measurable parameters within the structure (usually stress level or total deflection) and proceeds to check that this parametric value is not exceeded at normal working loads. This method, the allowable stress method, has the advantage of concentrating the mind of the designer upon the working condition of the structure and thus providing a realistic picture for the designer. The disadvantage of this system is that by concentrating on the expected it can lead the designer to ignore those events which occur as collapse is approached. This approach makes the assessment of risk of collapse very difficult.

The alternative system, limit state design, concentrates on the criterion which is being tested, in the case of structures either collapse or unserviceability. The designer must in this case work with values which are pessimistic, the amount of pessimism being relative to the amount of risk the designer is prepared to accept. In this case the model is less useful as an indicator of the normal situation within the structure, but the designer is able much more easily to assess risk than he could when using an allowable stress design method. If risk can easily be assessed, then so can its cost and the relative cost of its avoidance or reduction.

The design philosophy of the project planner does not approach the sophistication of his structural colleague. The original PERT program did allow the assessment of risk but it has seldom been used for this purpose in the UK construction industry. In general planners have been satisfied to work with the most likely values of the various parameters which concern them (analogous to the allowable stress method of structural design) together with the criterion of failure, the specified completion date of the project perhaps (analogous to limit state design). This mixed philosophy of design has not improved the usefulness of the results of the planners' work for the support of design decisions especially

## THE DESIGN CYCLE

where the planners have, intuitively, blurred the edges of the methods by introducing their own safety factors. There is clearly a case for considering both the definition of success and also the assessment of risk in project method design.

The third process in the design cycle is the making of the modifications shown to be necessary by the test. It is in this third process that the use of mathematical models yields its advantages. In either structural design or project method design a mathematical model can be changed quickly and easily and having been changed can rapidly be reanalysed. The designer will have many sources of information which will help him to decide the type and scope of the appropriate alteration; these will include his own experience of past design, the accumulated wisdom of his profession, and the iteration path of the present design: if he is to modify quickly and efficiently he must be helped to use these data sources. One way of doing this is to ensure that the information produced by the model is attractively presented in a form that makes the data easy to assimilate. In addition to this the design cycle must be fast enough to enable the planner to remember the detail of what has gone before (including the reasons for his actions) and also to allow him to use small iteration steps within the design cycle.

These needs of the designer have led to the present concentration of programmers on programs which allow designers to use the computer interactively. In most engineering disciplines this interactive use of the computer relies heavily on graphics and there is a strong case for trying a similar approach in the related field of project planning.

Although there are clear areas of similarity between structural design and project planning, and the consideration of the methods used in the first can point us to lines of possible development in the second, differences do exist. The identity between the two becomes uneasy when design philosophy is being discussed. It seems that in the case of structural design the limit state which is the basis of design, the collapse of the structure, is so horrifying that it is an absolute prohibition – it must not occur. By contrast the limit state of the project planner, the late completion of the project or the wastage of resources, is merely undesirable.

The debates which have accompanied the widespread introduction of limit state design into structural design codes, and the uneasiness about the unthinking use of cost–benefit analysis, have shown this dichotomy to be false. Structural failure is not prohibited, it is very undesirable and very expensive both in terms of the cost of replacement, the cost of human life, and the cost of lost pride.

Absolute safety is a fiction, almost absolute safety is usually extremely expensive, and the designer here as elsewhere must balance cost against benefit. The apparent difference between the design disciplines is merely a difference in the cost of failure. If, for example, human life depended on the timely completion of a project, the pressures on the planner would be exactly the same as those imposed by society on his structural colleague and project planning would be obliged to adopt a similar design philosophy. Usually the cost of planning failure is not of public concern and such pressures are not publicly applied, but this does not imply that the costs are negligible or that the examination of suitable design philosophies will not be useful.

## Pointers to change

This comparison of the design function in the fields of project planning and structural design has suggested several areas of useful development which will be examined elsewhere in this book. These areas are:

**(a)** The awareness by the user of project planning systems that he is dealing with a model and not with reality. The model he constructs must be appropriate to the task in hand and will not contain all the detail which would be required to represent reality. The designer of these systems should therefore be able to match the accuracy of the model, the cost of its operation, and the detail of the results with the needs of the user. The user of such systems should also be aware of the limitations of model testing and not use the model inappropriately.

**(b)** There is an evident need for an awareness by the planner of projects of the criteria of success towards which he is working. Such a reappraisal of design philosophy will only be useful if tools are available to help the planner to assess risk. Tools capable of the assessment of risk have been available for some time but have been little used in the construction industry; however, the latest generation of computer hardware has the power to do the necessarily long calculations quickly and so to make possible the use of these tools within an interactive design environment.

**(c)** Planning has been shown to be cyclical. If the planner is to be creative he must be given the same type of support which, it is increasingly being seen, is necessary for all designers – that is, computer aid which is interactive. Planning uses large volumes of data and this data can be quickly assimilated most easily using graphical input and output; a project planning system should therefore perhaps be based on interactive graphics.

This analysis has considered only a two-dimensional model of the design process. The possibility of a third dimension must be considered before a euphoric start is made on the redesign of project planning systems.

## Analytical design – a hierarchy of detail

The two-dimensional model of the design process adequately describes the action of the designer when the test is unsuccessful, but does not show his actions if the test succeeds. In order to do this the designer must introduce a third dimension to his design procedure. This is illustrated by Figure 2.3.

As the figure shows, a successful test does not now imply that the design process is complete, if the designer is satisfied with the test he proceeds to add detail to the design and proceeds in this way down a hierarchy of design

## ANALYTICAL DESIGN

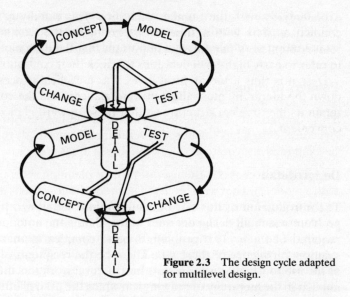

**Figure 2.3** The design cycle adapted for multilevel design.

decisions until he achieves the detail which is appropriate for his purposes. This hierarchy of decisions is the means by which the complexity of the decisions is reduced in all design disciplines, enabling decisions to be made rationally. For example, the structural engineer, asked to design a bridge across the Humber, postpones decisions concerning the detailed fabrication of the deck until he has decided whether he is going to build a suspension bridge or a stayed girder. Decisions made at this stage are provisional, and he may return to reconsider them in the light of information gained at lower levels. Progress in design is impossible unless some form of hierarchical strategy is adopted.

The complexity of designs built in this hierarchical way is dependent on the number of layers included. Figure 2.4 shows how the complexity grows

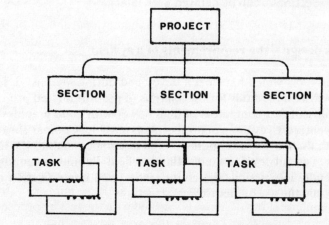

**Figure 2.4** The growth of complexity in hierarchies.

15

exponentially with the number of layers. This rapidly growing complexity, coupled as it is with a time commitment by the designer, makes radical reassessment very painful; it is important that the designer frequently returns to refer to early high-level decisions to check their continuing validity.

Design is thus a hierarchical as well as a cyclical process. The movement down the hierarchy must allow upward glances, and the complex whole must retain its hierarchical structure so that these upward glances are possible and convenient.

**Design detail**

The introduction of this third dimension into the design process highlights a problem which all designers must solve, that is the amount of detail which is required of them. This problem is quite complex in that it requires for its solution knowledge of those who are to be the recipients of the work and also of the use to which it is to be put. In structural work this difference in detail is implicit in the hierarchy of drawings on which the design is described. In project planning it is implied by the differing detail of the Gantt charts found at different levels within organisations.

It was argued in the first chapter that although the management of projects does require fine detail, much of this detail relies on data which are not available at the start of the project. Decisions regarding detail must be added to the plans at a later stage when all the necessary data are known and thus the shape of the project information content will approximate to the information wedge. This procedure clearly is supported by a hierarchy of decisions: the planner concentrates his efforts during the pre-project planning stage on the decisions which he can support with the data to hand and which help him with the task in hand, be it cost estimation or material ordering; the lower layer decisions – which gang will do a particular job or the precise order of erection of the precast sections – can be delayed until later.

**Planning as design – the requirements of a system**

The planning tools which have been developed during the past decade have admirably provided methods for the *analysis* of pre-formulated project plans. The availability of fast computers of high power with good graphics facilities has made the interactive *design* of project plans feasible. Comparison of project planning with the more conventional design disciplines enables the identification of areas of useful development in this new field. Paramount among these is fast and friendly interaction, the consideration and quantification of design uncertainty, and the use of hierarchical structures.

# 3
# THE NETWORK MODEL

**Models**

In order to do the thinking necessary to make decisions about a project, we must form a mental representation of it which must normally be a simplification of reality. The degree of simplification will be dependent upon the use to which we put our representation; we need a very much more complex picture of our spouse than of the man in the paper shop because our interaction is more complex and important with the former than with the latter.

The previous chapter showed that in the design process a representation is required. This representation, which we called a model, could be physical, a scale or a distorted model of the projected design, or mathematical, a series of equations representing in mathematical form the way in which the projected product or design will act. The model will not contain all the complexity of reality, for in order to do this the design model would cost as much as the prototype and design interaction would be slow and extremely expensive; it is so built that the designer can use it to show how particular features of the final design act in particular circumstances. Thus a designer responsible for the towers of a suspension bridge uses one model for his aerodynamic studies (probably a physical model) and a second for his structural analysis (usually a mathematical model); neither completely represent reality, but each provides information about a particular feature of the design.

Thus for the design of a project we are looking for a representation of the project which can be used to demonstrate particular features of the design and which can be handled and changed conveniently and cheaply. The modelling system usually used is that of representing the project as a network.

**Network models**

A project exists in time and thus is a sequence of events; the project can be said to be complete when all these events have taken place. The events are not independent of one another, they cannot all take place on the first (and therefore the last) day of the project; some events can only take place after

## THE NETWORK MODEL

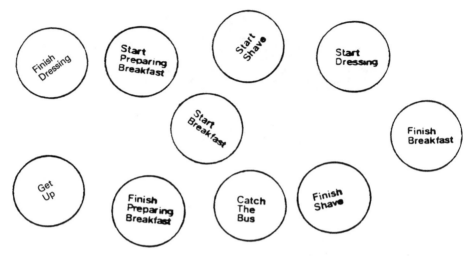

**Figure 3.1** The freely floating activities.

others have occurred, thus the sequence and timing of the events is dictated by logic. Figure 3.1 shows a number of events all of which must take place if I am to arrive at work in the morning. These events are shown as drifting freely in space like balloons, but in fact they are firmly tied together, for I cannot start to shave until I have got up; I cannot finish shaving until two or three minutes after starting, and so on. The complete mooring system for those balloons is shown in Figure 3.2.

It will be noticed from Figure 3.2 that some of the strings between balloons are of unspecified length while some have a minimum length specified for them.

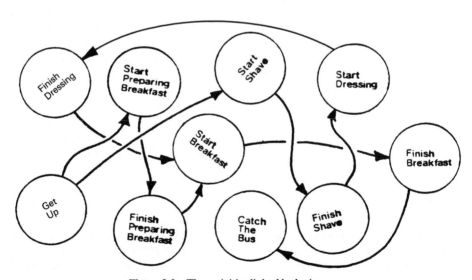

**Figure 3.2** The activities linked by logic.

# CONVENTIONAL NETWORK REPRESENTATIONS

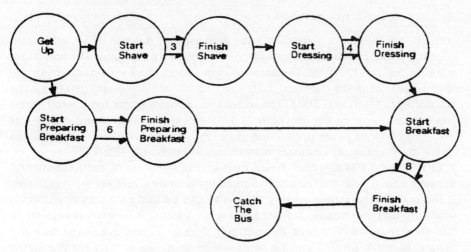

**Figure 3.3** The element network.

Thus the string between 'get up' and 'start shaving' could be of zero length or could, if I was particularly sleepy, be many minutes long. The string between start shaving and finish shaving cannot, however, be less than two or three minutes long. The start and finish of the activity 'shave' must be separated by at least the time required for me to carry out the operation, the duration of the activity.

It will also be noticed that this representation of the project entitled 'get to work' consists of a network of strings connecting together events, and that each activity (with the exception of the start activity 'get up' and the finish activity 'get the bus') has two events; that is, start the activity and finish the activity. Thus our model of the project has taken the form of a network based on logic which ties together the events which comprise the project, and, in a somewhat tidier form, can be drawn as Figure 3.3. In this figure the minimum times of activities are shown between the double arrows.

## The conventional network representations

Figure 3.3 is a rigorous representation of the network of events which leads to me catching the morning bus. It can be simplified in one of two ways: by combining the events which can in theory be carried out concurrently (finish shaving and start dressing can, for example, be seen as one event); or by representing as one node the two events which, together with the arrow between them, represent an activity – start shaving and finish shaving, for example. Both these systems of representation of networks are used in project management: the first is known as **activity on the arrow**, examples being the well-known CPM and PERT programs; the second is known as **activity at the node**, and PDM is an example of this.

## THE NETWORK MODEL

*Activity on the arrow networks*

Activity on the arrow networks were first used in the late 1950s when two research teams coincidentally invented very similar techniques. The first team, working for the Du Pont Corporation, was concerned with minimising the 'down-time' of petro-chemical plants during maintenance and produced the Critical Path Method (CPM). The second team, working for the United States Navy, was directing the research and development programme which was eventually to lead to the production of the Polaris missile; their technique was called the Programme Evaluation and Review Technique (PERT).

PERT and CPM are each based on the simplification of the fundamental network which occurs if the events linked by 'zero time' arrows are combined. If this is done the arrows joining the nodes can be said to represent activities while the nodes continue to represent events, although now the events are in fact combinations of two or more simple events. Figure 3.4 shows how the getting to work project can be represented using an activity on the arrow network.

It is clear that Figure 3.4 represents a considerable simplification of Figure 3.3 and that the diagram is easier to understand (it is immediately apparent, for example, that the logic of the diagram depends upon the existence of somebody to prepare my breakfast while I shave and dress). The simplification has, however, been bought at the price of some ambiguity, for the nodes now represent more than one event, two in the case of B and D (finish breakfast and catch the bus), three in the case of A and C (finish dressing, finish preparing breakfast, and start breakfast). If, as is sometimes the case, it is necessary to limit the events at the node, then the 'zero time' arrows of the fundamental network must be reintroduced. Thus, if for example the preparer of the breakfast wishes to dress immediately after breakfast was prepared, merely adding an arrow coming from C and entitled '2nd dressing' does not represent the true situation, for this activity is dependent only on 'prepare breakfast' and is independent of the activity B–C 'dress'. Fully to represent the situation the 'zero time' arrows (called dummy arrows), must be shown in the network as

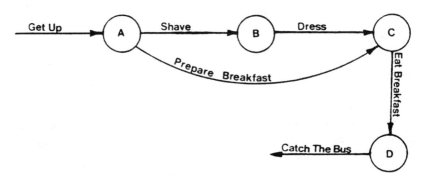

**Figure 3.4** The network represented as 'activity on the arrow'.

## CONVENTIONAL NETWORK REPRESENTATIONS

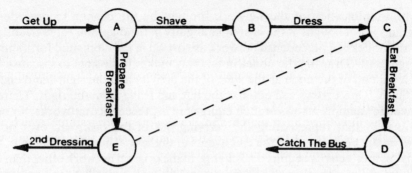

**Figure 3.5** The inclusion of the necessary 'dummy' arrow.

illustrated by Figure 3.5. As the complexity of networks increases, so does the proportion of dummy arrows in the network, robbing the diagram of some of the simplicity which was won by combining events.

## Activity at the node networks

The second possibility for the simplification of the fundamental network is to combine the two events which mark the beginning and end of an activity, and the arrow which represents the activity itself. This combination leads to a technique known as the Precedence Diagram Method (PDM) which was introduced in the 1960s. PDM is used in the construction industry to an increasing extent, although the ten-year lead by the activity on the arrow diagrams has built a considerable 'brand loyalty' for the older techniques. A precedence diagram for the network shown in Figure 3.5 is shown in Figure 3.6.

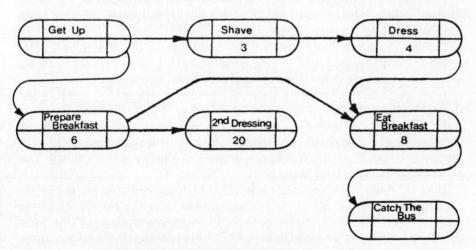

**Figure 3.6** The network in precedence form.

## THE NETWORK MODEL

The simplification of the network brought by the change to precedence diagrams is not bought at the cost of ambiguity in the diagram, for conventionally the nodes of a precedence network are drawn in the elongated form shown in Figure 3.6. Drawing the nodes in this way makes it possible to visualise the left-hand end of the node as the start of the activity and the right-hand end as the finish. Thus arrows will enter at the left and leave from the right. There is no need for dummy arrows or their equivalent in precedence networks, for only one arrow, that representing the carrying out of the task, can ever occur between the beginning and the end events of the task. Thus the addition of the '2nd dressing' activity required no change to the original network other than the addition of the activity itself and the arrow linking it logically to the end of the activity 'prepare breakfast'.

### *A comparison of the two networking systems*

Although CPM and PERT have established a firm hold within the construction industry planning environment, there are strong arguments in favour of a change to the use of PDM. These arguments derive from the lack of ambiguity in PDM and from the relative ease of the use of PDM within computer programs.

It has been pointed out above that the CPM simplification of the original network introduces ambiguities. These ambiguities can be overcome by the use of dummy arrows but this destroys some of the simplifications and introduces a potentially dangerous dual status of arrows, some representing both logic and activities while others, the dummies, representing logic alone. While the presence of dummy arrows may not be troublesome in simple networks, in complex networks they will dominate the diagram. The precedence diagram, on the other hand, contains only one type of arrow, that representing the logical link between activities; the activity arrow has been combined into the node. The ambiguities inherent in CPM and PERT will again become apparent when the analysis of the diagrams is attempted.

The specification of the network for computer use is much simpler if PDM rather than CPM is used. In CPM the nodes represent rather ephemeral events and the activities are defined as linking these. Thus the operator of a computer program must first draw out the network, numbering the nodes, and then use this network during the input stage of the program. With PDM this is not so. In order completely to define the network the user must merely have a list of activities and be in a position to specify which activities are dependent on or precedent to other activities. No prior drawing of a network is necessary. This point will be pursued later in the book.

Thus although there are two networking systems available for use in the construction industry, all the advantages, other than the familiarity of the industry with CPM, favour a change to the widespread adoption of PDM. Most of the subsequent discussion in this book will be concentrated, therefore, on the use of precedence diagrams as the most useful model of projects.

## The analysis of simple networks

It has been argued above that a project is a series of events and the model which has been developed is such that the sequence of the events can be predicted mathematically by forming the events into a network. The next concern is to develop a mathematical procedure whereby the network can be analysed.

All the events in the network could take place at any time but, given a particular start date of the project, there is a limit to how early the event can take place. Thus in the simple example, if I get up at 7.30 a.m. I cannot possibly finish shaving before 7.33 a.m. The first part of the analysis of networks is concerned with the identification of this earliest possible time for each event in the network; this operation is known, for reasons which will become obvious, as the forward pass.

If I carry out the forward pass for the whole network I will obtain an earliest possible time for the last event 'catch the bus'. (In our case, as we shall see, this time is 7.45 a.m.), and if my aim is to complete the whole project in the shortest possible time, then this early time for the final event will also be the latest permissible time. If I wish to catch the 7.45 a.m. bus then I must start my breakfast not later than 7.38 a.m. The calculation of these latest permissible times forms the second part of the analysis and is called the backward pass.

### *The forward pass through the network*

In order to calculate the earliest possible time of an event all the events which must, following the logic of the project, precede it must be considered and allowance must be made for the durations which are assigned to any of the network arrows. In the case of the precedence network, where all the arrows have zero duration, the start time of an activity will be set to the latest of the finish times of the precedent activities.

As this process continually looks backward to the precedents, it is most efficient to carry out the calculations systematically, working from the start event, which has no precedents, to the end event which has every event in the network as its precedent. In this way the calculation is progressive and cumulative, an event is analysed only when all its precedents have been themselves assigned an early time. It is because the calculation works progressively forward through the network that it is called the forward pass.

By convention the upper part of the precedence network node is used to show the early dates, that is the upper left to show the early start of an activity, the upper right to show the early finish. Figure 3.7 shows our simple network with the early event times inserted in the appropriate places. The calculation proceeds in an orderly fashion from the value of 7.30 a.m. for 'get up' setting each start time to the latest precedent finish time and each finish time to the start time plus the duration and the values are the earliest possible times for the various events.

## THE NETWORK MODEL

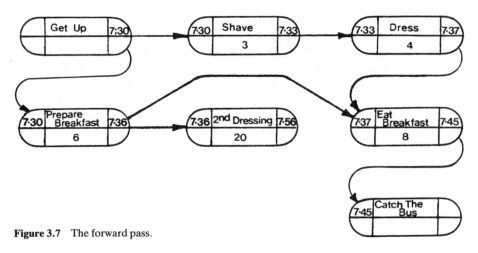

**Figure 3.7** The forward pass.

## The backward pass through the network

If the end event of the project is to take place at the earliest possible time, then the times of all the other events in the network which affect it have a time which they must not exceed. To calculate this latest permissible time a process similar to that developed for the forward pass is required. In this case, however, the analysis is concerned not with the events between the start of the project and the event being considered, but with those between the event and the end of the project. Again the analysis of the precedence network is simple, the finish events of activities are set to the earliest of the start times of the dependent activities, and the start time is made equal to the finish time minus the duration.

The convention is to show these 'late' dates in the lower part of the precedence network node, again the left representing the start and the right representing the finish. Figure 3.8 shows the example network with the late

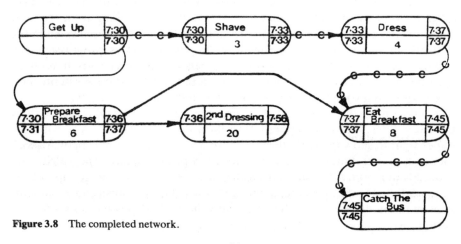

**Figure 3.8** The completed network.

times inserted. Once again the calculation proceeds in an orderly fashion, but in this case the progression is backwards from 'catch the bus' which has no dependent activities to 'get up', and hence the name 'backward pass'.

It should be noted that in this network there are two end events, 'catch the bus' and the finish of '2nd dressing'. This latter activity is not a precedent of 'catch the bus' and thus has no late times. Some computer programs prohibit this use of multiple finish (or start) activities, but as this example shows there is no reason in logic why there should be a single start or finish event.

## Criticality and float

The forward and backward pass through the network produce for each event a range of times at which the event can take place without jeopardising the completion date of the project. The limits of this range are the earliest possible event time, produced by the forward pass procedure, and the latest permissible time, the product of the backward pass. In some cases these two event times will coincide, and the earliest possible time will also be the latest permissible time, and then any slip of the event to a time later than the earliest possible time will delay the completion of the project. These activities are known as critical activities, and the path through the network containing them is called the critical path. The critical path through the example network is marked with a series of Cs in Figure 3.8.

Events away from the critical path have a range of possible times, and this range is known as float. Thus float, the difference between the early and late times of an event, is a measure of the tolerance within the network, and comparisons of the float of events can help the manager of a project to assess priority. In the example network the activity 'prepare breakfast' has a float of one minute. Inspection of the network shows that three options are open to the preparer of breakfast, he can either delay the start of breakfast by one minute (i.e. stay in bed!) or he can take one minute longer in the preparation, or he can finish breakfast one minute early and allow the toast to go cold. The presence of float has created these options.

Although this is a trivial example, it is not difficult to see that in complex projects where, for example, expensive resources are to be shared, the ability to delay the start of one activity or, through the sharing of a common resource to extend its duration, can be of major importance to the manager. Criticality and float make such decisions possible.

## The network model

This chapter has introduced the network model and its analysis in the simplest possible form. Later chapters will show how the basic model can be elaborated to increase its realism and how, using computers, the analysis of network models and the presentation of the results of the analysis can be made such that they can form part of a continuous design operation.

# 4
# GRAPHICS AS AN AID TO INTERACTION

**The need for graphics**

Chapter 2 emphasised the need for fast interaction within the design process and, in establishing the similarity between project planning and product design, showed that this need also exists within project planning systems. This chapter will examine this need and the ways in which the high-resolution graphics now commonly becoming available on quite small computers can be used to meet it.

The analysis of project plans consists of three phases, the assembly of the data, the construction and analysis of the mathematical model of the project, and the assimilation of the results. During the past 20 years, as computers have become more easily accessible, there has been an increase in the proportion of these tasks which are carried out using them. Figure 4.1 shows how the introduction of computers into the various tasks might affect the total time required for one cycle of the planning process for a one hundred activity network.

Although Figure 4.1 is based on purely notional figures the conclusions to be drawn are obvious. The first conclusion, of course, is that merely to use the computer to carry out the arithmetical analysis of a previously constructed network is not efficient. It is possible that using the computer in this way will result in loss of time rather than saving. The time saved due to the almost instantaneous analysis is more than offset by the time required to type the data into the computer. This loss is doubly important, for not only is time lost but also the time spent on analysis, which most planners find enjoyable, is replaced by time spent on typing, which few enjoy. It is not surprising that programs designed to do only the analysis of previously drawn networks are not popular. It should be said of course that the time saved by these simple programs is more significant if the data input is to be used more than once (either as part of an iterative design process or as a way of updating); however, few of the simple analysis programs are such as to allow this use.

The second conclusion to be drawn is that the second stage of computer

# THE NEED FOR GRAPHICS

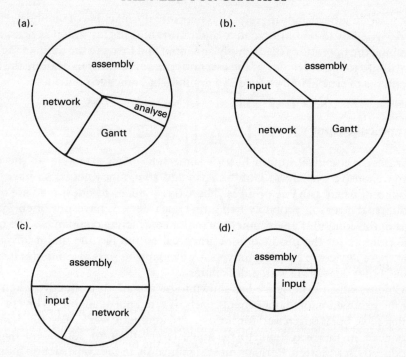

**Figure 4.1** The evolution of computer-aided project planning; (a) manual analysis – 7½ hours; (b) simple computer analysis – 8 hours; (c) computer analysis with graphics output – 6 hours; (d) precedence network computer analysis – 4 hours.

application, the production of graphical output, brings with it major time saving. Even the one-run example shown in Figure 4.1 shows a saving in time if the results of the analysis are automatically produced as charts rather than as columns of numbers which then have to be plotted. This saving becomes very large if there is the need and the capability to update plans and produce new output. There may, however, be a trade-off problem here in that the quality of the charts may not be as good as that produced by a draughtsman.

A third conclusion may be drawn from the fourth diagram, which shows how, by combining the technique of precedence networks with computer analysis and automatic graphics, the time required for analysis can be further and significantly reduced. The use of precedence networks removes the need for the prior drawing of the network, necessary in the critical path method in order to ascertain node numbers, and so to define activities. The removal of this time-consuming operation is welcome, but again there is a possible trade-off problem, the planner does not now have access to a network to guide him in his design iterations and this may inhibit his further performance.

At the end of this evolution of method we are left with an operation which consists of the assembly of data and its input into the machine. Data assembly is an activity which relies heavily on the experience and flair of the planner; data entry, while superficially merely the typing into the machine of numbers, can be so designed as to be both creative and enjoyable. In this way it is possible to combine the operations of assembly and input and to make the input of the data contribute to the efficiency of the assembly phase. If this is to be done the form of entry is of crucial importance and graphics has much to contribute.

**Graphical output**

Hard copy graphical output has for some time been available on the more comprehensive packages: Gantt charts and sometimes networks have been produced, usually on line printers. There has not been a widespread use of pen plotters, the reason probably being that such devices have not often formed part of the computer environment of major contracting companies. The use of line printers for the production of graphical output has the major advantage that these devices are almost universally available as part of computer installations, but it does bring with it difficulties.

Colour, which experience has shown to be remarkably effective when used for the production of management charts, is often not available on line printers which have been purchased for normal alphanumeric work; also the diagrams produced in this way tend to be much larger than an equivalent diagram produced on a plotter. Perhaps most significantly in the construction industry, where there seems to be a particular phobia about computer use, the output produced by means of line printers looks very different from that produced by draughtsmen and is therefore less easily accepted. Figure 4.2 shows a rather startling example of this. Experience of attempts to introduce planning systems into companies shows that the appearance of output is of vital importance.

For these reasons it seems that ideally line printers should be used as secondary options only and that pen plotters or one of the printers especially designed for graphical use should be the peripherals around which computer-aided planning and control systems are built.

The difference between hard copy produced on a line printer and that produced on a plotter is demonstrated by Figures 4.2 and 4.3. Figure 4.2 shows a simple network diagram produced on a high speed line printer. The line printer, although having the advantage mentioned previously of ease of access, produces diagrams which are very large, for the size is dictated by the size of the typeface. For example, the node of a precedence network must be at least 13 letters by 5 letters. The resulting diagrams can be so large as to become unwieldy (one diagram produced in this way for a large civil engineering project with fifteen hundred activities measured 5 m × 3 m!), and diagrams such as this can easily become the source of jokes rather than of management aid. Diagrams produced in this way will normally be monochrome (although some programs make very effective use of colour printers).

By contrast Figure 4.3 shows a small part of a complex network drawn by a pen plotter. Size is now much less of a problem because the character size of the

# GRAPHICAL OUTPUT

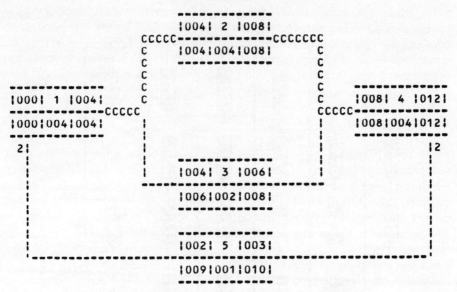

**Figure 4.2** A line printer produced network diagram.

plotter can be altered at will and very small legible characters can be produced. The diagrams are therefore easier to handle and, because they are composed of drawn lines, look much more familiar to those who are used to diagrams produced by draughtsmen. Colour is usually available on plotters and a diagram such as Figure 4.3 can show how effective a quite simple addition of colour can be. In this case the activities lying on the critical path can be drawn in red and the diagram, which ordinarily would be very difficult to follow, becomes relatively easy to understand. In the same way colour can be added to Gantt charts such as Figure 4.4, to show criticality or completion.

It has been shown earlier in this chapter that network analysis can be carried out without the prior drawing of the project network diagram, but that planning a project without the production of a network such as that illustrated in Figure 4.3 will be difficult if design iteration is to be carried out. Nevertheless most of the output of planning programs will be in the form of Gantt charts. This is the form in which the information is traditionally displayed to the workforce and the form with which the industry is familiar. Here again it is possible to use line printers (with the same difficulties of size and colour) and here again plotted output shows considerable advantages. Mention has already been made of Figure 4.3 which shows the advantages of these devices which are becoming increasingly cheap and robust. Figure 4.5 shows, at a much reduced scale for the purpose of this book, a plotted diagram including both a Gantt chart and a network diagram such as has been produced for one civil engineering company. This type of diagram has been very well received.

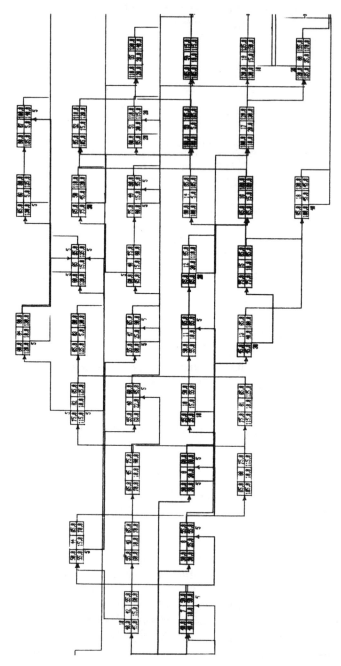

**Figure 4.3** A plotter-produced network diagram.

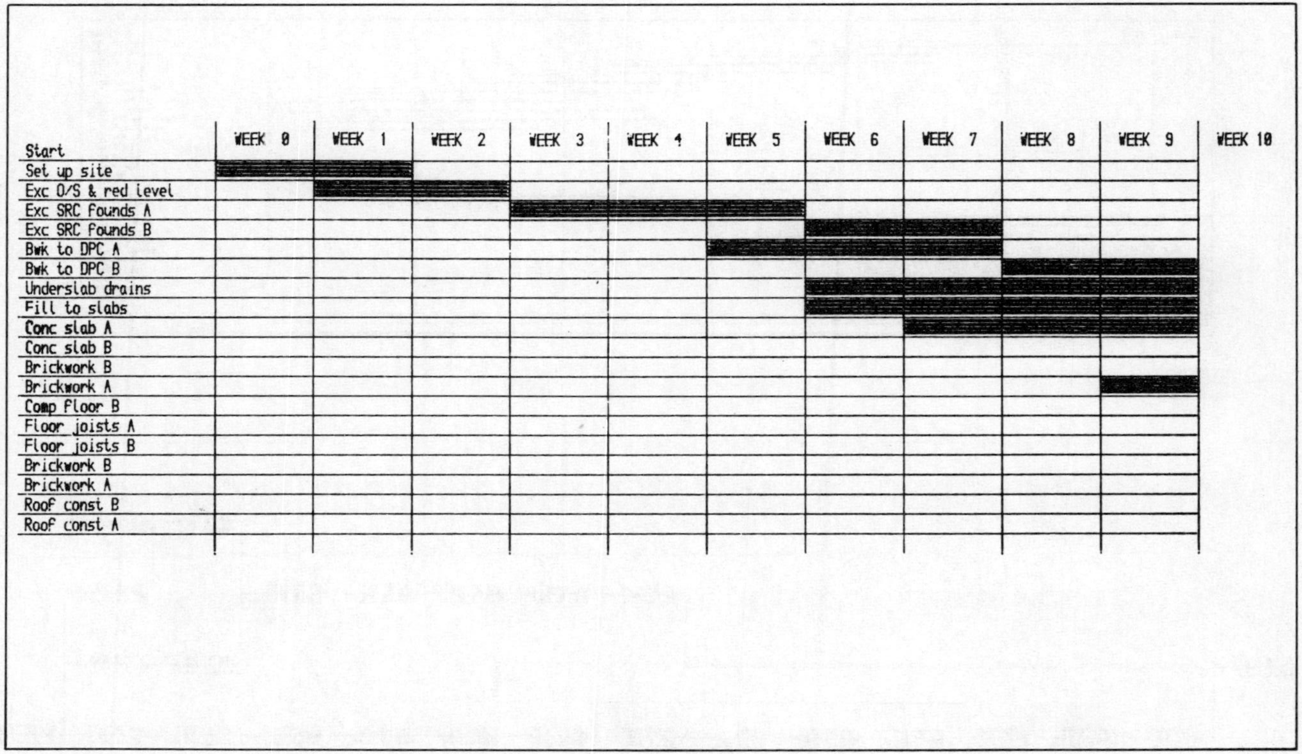


**Figure 4.4** A plotted Gantt chart.

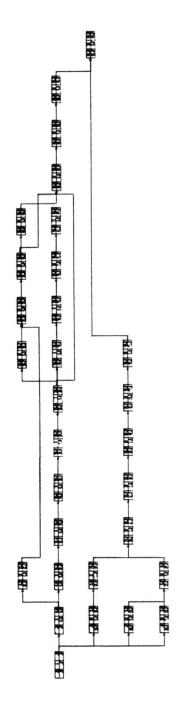
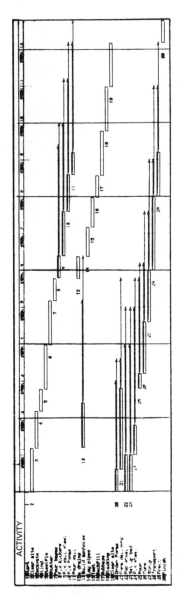

**Figure 4.5** A complete plotted output (much reduced).

## GRAPHICAL INPUT

In addition to hard copy graphics the possibility now exists for the production of graphics displays on inexpensive terminal screens. This enables the planner quickly to sample graphical output and therefore makes the interactive use of planning programs a real possibility. The size of such screens is, of course, limited and therefore in all but the most simple cases the output must be displayed in sections or, better, scrolled, but this presents no major programming difficulty. It has been found that at the terminal as in hard copy the use of colour contributes strongly to the speed and accuracy of assimilation of information presented to the planner for action and thus to the quality of decisions he makes.

Graphical output from analysis programs is being seen to be invaluable in all engineering disciplines; project planning is no exception to this and is perhaps, because of the tradition of the industry and the relative simplicity of its basic chart, a prime candidate for its introduction.

## Graphical input

The input of data for planning programs is tedious and, as such, error prone. Many planning packages, recognising this, have incorporated attractive and user-friendly input formats which reduce the problem. The problem remains, however, for a considerable volume of data must be provided for even a modest network analysis.

If the method of analysis is to be CPM or one of its derivatives, then a network must be drawn in order to define the activity numbers before data input can begin. This is a time consuming occupation which requires considerable care if errors are to be avoided. Transmission errors are likely when this information is taken from the diagram to the computer, for the numbers bear no relation to the planner's intuitively known model of the project. This difficulty is the major inherent disadvantage in the use of activity on the arrow networks for computer analysis.

The use of precedence networks removes the need for the prior drawing of the network, but does not remove the difficulty of data input. Input data must now contain, not the bracketing node numbers of the activity, but a full list of the precedent activities. This list of numbers must be assembled by the planner from a full activity list and again there is a high probability of error as the data are transferred from an external list to the computer data store. An error in this operation is probably more difficult to detect than the equivalent error in CPM, and can often result in a loop in the logic of the network.

Graphical means of input can reduce the risk of error in three ways. First, data input by graphical means is a more pleasant occupation than is the typing of columns of figures, and if the planner is less bored he will make fewer mistakes. Secondly, graphical input is so much easier to check than numerical input that it is virtually self-checking. Thirdly, graphical input can be so structured as to be a part of the data assembly process; it is therefore possible

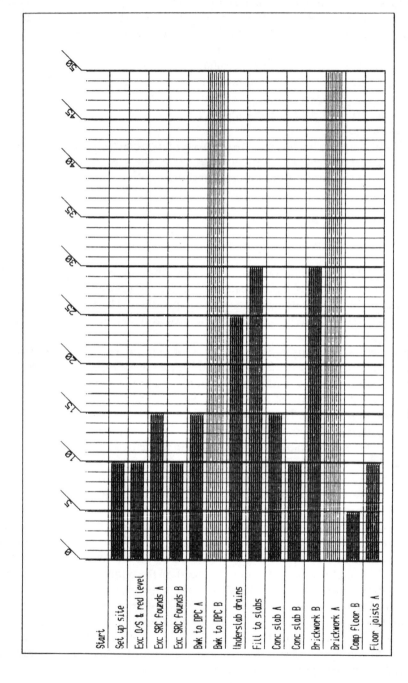

**Figure 4.6** The duration histogram.

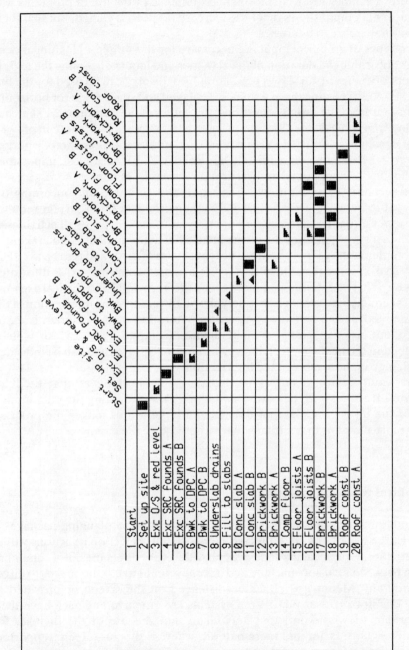

**Figure 4.7** The dependency matrix.

to integrate the creative data assembly part with the tedious data input part, and the latter becomes lost in the former. Experience of the use of programs with graphical data input shows that little increase in speed is gained, but that there is a major increase in interest.

Two types of graphical input are necessary for the simplest planning model: input concerning the duration of the activities and input concerning the logic of the network. Figure 4.6 shows how a graphical display can be used for the first of these. A chart listing the activities is displayed and space is left for horizontal bars to represent the activity durations; the values of the durations are sketched in using a graphical input device (a light pen, a joystick-driven cursor, or a cursor driven by a digitiser pad or a mouse). Durations are easily crosschecked and gross error is so immediately noticeable as to be almost impossible. Scrolling in both horizontal and vertical directions is, of course, necessary.

The second type of input which is required is information concerning the relationship between activities. An alphanumeric data input system requires the planner either to have a superb memory or to make a list of all activities so that he can specify the source of the links by activity number. The making of such supplementary lists is clearly a potent source of error and is to be avoided. A way to overcome this difficulty is to use a 'dependency matrix' as illustrated in Figure 4.7. The activities are listed along the horizontal and vertical axes and a logical link between them is uniquely shown by marking the intersection of the lines defined by the activities. Again the alteration of the diagram is easily carried out using a graphics input device. Complex links (which will be described in the next chapter), if incorporated in the program, can be indicated by a change in the symbol at the intersection.

This input device is not as immediately self-checking as that used for duration. It shares with it, however, the property of making the data input a part of the decision making of data assembly, and thus makes the program much more pleasant to use than would otherwise be the case.

**Graphical program direction**

An additional way in which graphics can contribute to planning packages is through the use of graphics directed menus for the selection of actions within the program. Graphical input devices provide a very direct means of communication between man and machine and it is convenient to use this means to direct the program. Menus can either be displayed on the screen or provided as 'paste-ons' for digitiser pads. Both systems are effective and each have their supporters: the paste-on supporters claim that the use of the digitiser for direction is both faster and more natural, whereas the 'on-screen' supporters point to the foolproof nature of their system. Either of these systems can be used to prevent the frustrating question-time sessions which were necessary for the direction of non-graphical interactive programs.

## The use of graphics

Experience of the use of the programs described in this book has shown that the use of graphics for the handling of output data from network analyses is both efficient and well received. Although PDM theoretically removes the need for the drawing of networks, practice suggests that for all but the most simple projects management will benefit if a plotted network is available. The use of colour is an advantage for both networks and Gantt charts.

There is less experience of the use of graphical input, but such as there is supports its use. It seems that input in this form is less error prone than its numerical equivalent, and it brings the major advantage of being pleasant to use. Graphical input can, therefore, allow network analysis to realise its potential by transforming it into a cheap and responsive management tool.

# 5
# SOME NON-GRAPHICAL PLANNING AIDS

**Improvements to realism**

The use of graphics can enormously improve the friendliness of computer packages and thus improve the quality of the interaction between man and machine. This improvement can be increased by making the model of the project on which the planner is working as realistic as possible. The aim of the developer of all interactive design programs should be to reduce to a minimum the mental gymnastics which the designer has to perform as he takes the presented results and assimilates them prior to deciding his next action. A word of warning should be sounded, however, for as pointed out in Chapter 2, a model should be suitable for the purpose in hand and for many purposes the sophistications to be described here may not all be appropriate.

In this chapter improvements to realism which do not rely on graphics will be described. Most of the techniques described are possible in the non-computer environment when used individually. Their use in combination would face the manual analyst with a complexity such that at the least the analysis would lose the elegance which makes the hand calculation of networks pleasant and that in the worst case would force the planner to make mistakes and also give him a headache!

Five techniques will be described: complex links, project closedown, window times, varying production rates, and calendar dates.

**Complex links**

One of the advantages of the precedence diagram method is that the specification of links which are more complex than the direct link described in Chapter 3 is allowed. These complex links, which, together with their equivalents in the fundamental network model, are shown in Figure 5.1, enable the planner to describe relationships which can only otherwise be described by the splitting of

## COMPLEX LINKS

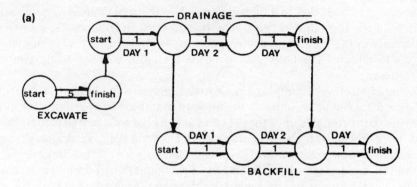

THE FUNDAMENTAL NETWORK

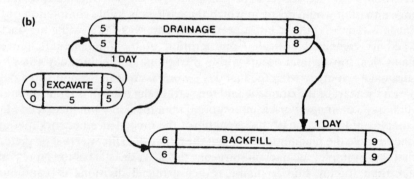

THE PRECEDENCE NETWORK

**Figure 5.1** Complex links; (a) the fundamental network; and (b) the precedence network.

activities and the resulting increase in complexity. The complex links can be 'leads', which specify that the start of the activity is in some way dependent on the start of the precedent, or 'lags', in which case the finish of the activity and the finish of the precedent are tied together in some way.

Complex links allow the planner to use larger activities than would be possible otherwise, and so reduce the complexity of networks both for the analyst and for the user of the results. Their use should be limited, however, for it is tempting for the planner to use them to form 'omnibus' activities which contain a large number of events and which, because they are amorphous, make detailed control impossible.

Complex links are in fact evidence of the hierarchical plan structure briefly described in Chapter 2, for they imply a series of low level plans within the stated activities. The development of hierarchical planning systems, which is discussed later in this book, should theoretically render them redundant.

Experience shows that in a typical network in the civil engineering industry 10–15% of the links in a network may be complex. In some cases, however, particularly in the building industry, wide use is made of very large activities and up to 80% of the links may be complex (with corresponding control difficulties).

The incorporation of complex links into the analysis is not difficult. Referring to Figure 5.1 it can be seen that in the forward pass the start of activity 'backfill' is tied not to the completion of the previous activity but to its partial completion. Thus 'backfill' can start after one day's work has been done on 'drainage', and thus its start is tied to the start of 'drainage'. Similarly for the 'lag' link tying the start of the last day's work on 'backfill' to the completion of drainage. In each case the usual forward and backward progression through the network is used as the basis of calculation.

Two points should be noted. If the lag link in Figure 5.1 were not present there would be no tie between the finish dates of the activities and the rather strange situation would occur in which the earliest possible completion of the drainage was day 8, but the latest permissible date was day 9, while the start of the activity remained critical. Consideration of the fundamental network explains this, for the float occurs within the second or the third day's work on the drainage activity leaving the first day's work as critical. Secondly, the delay between the start of the drainage and the start of the backfill is not absolute. If the first day's drainage work is interrupted, then the delay will be extended.

Complex links are one of the strengths of the precedence network method, for they enable the planner to represent the network in the way that he pictures it (that, for example, the backfill starts one day after the start of the pipelaying) rather than forcing him to make rather artificial divisions in continuous operations. While they should be available as options in all PDM programs, their use should be carefully monitored if difficulties in the control phase of the project are to be avoided.

## Project closedown

It is very unusual for projects to continue without a break other than the usual stoppages for weekends; even the busiest project has to accommodate national holidays. When these stoppages occur they obviously lengthen the duration of the current activities. One way of dealing with this is to construct plans in terms of project days, the project day number being equal to the number of working days since the beginning of the project. This approach is widely used in the non-computer environment but has disadvantages. The major disadvantage is that the construction of a separate project calendar divorces the project activities from the outside world and makes interrelation between them more difficult for the planner and the manager. A second problem caused by this approach is that such an analysis method cannot handle those activities which are unaffected by closedown; examples of these include the curing of concrete and the consolidation of an embankment under surcharge, those which continue whether there are men on site or not.

An alternative to the project calendar approach which avoids these difficulties but at a cost of extending the analysis time, is to consider the effect of closedown within the analysis itself. As the duration of each activity is being incorporated into the network (i.e. at the calculation of finish times in the forward pass and start times in the backward pass) the location of the activity in time is checked to see if it coincides with a closedown period. If it does, then the duration is extended by a time equal to the length of the closedown.

## Window times

A very frequent constraint on the planner of the projects is the requirement that an activity should fall within the specified period. Examples of this are the 'possession times' provided by railways for the carrying out of work near their tracks and the 'weather windows' which have dictated the launch and positioning of offshore structures for the oil industry. This type of constraint is easily built into the basic analysis method for the forward and backward pass merely by setting the event times prior to analysis for the forward and backward passes respectively to the start and finish of the time window for the activity concerned, and then only altering the event times if they are to be increased (in the forward pass) or to be reduced (in the backward pass).

Experience shows that very rarely are project plans free from constraints of this type. The absence of the facility automatically to include them in the program seriously inhibits the usefulness of planning programs, but their inclusion is both easy and cheap.

## Calendar dates

The tying of the project events to calendar dates has already been shown to be important. The major advantage of the provision of calendar dates is that it removes the need for the planner to carry two calendars in his head, making the use of output much more easy. It is particularly valuable where the project is tied to external events as when, for example, window times are being used. The provision of calendar dates is one of the essential features of a useful planning program.

Although the calendar date is of great help to the planner, this help is necessary only at the output stage. There is, therefore, merit in delaying consideration of calendar dates until output data are being presented and in using a project calendar within the analysis.

## SOME NON-GRAPHICAL PLANNING AIDS

### Seasonal variations in production

The construction industry in the UK is particularly sensitive to adverse weather conditions, and this sensitivity must be reflected by the planner as he positions activities in time and estimates activity durations using forecasts of production. Where this adjustment is made it is made after the analysis of the network by extending those activities which have been shown to occur during the winter months and reducing the times of those which fall in the summer. Many contractors have curves of productivity which act as guides for the planner and the site manager; one such is shown in Figure 5.2. These curves correct not only for the effects of bad weather but also for the shorter working hours of the northern winter and the effect on the morale of the labour force due to working in unpleasant conditions.

A computer technique which allows for the seasonal variations to production is obviously of benefit, for all activities can then be estimated as if the work were to be carried out at those times of the year (April and September) when an average production can be assumed. Like the provision of window times, this facility is very easily built into the analysis method.

The ability to adjust times automatically is advantageous not only to the planner of projects but also to those responsible for the control phase and, perhaps most lucratively, those concerned with the costing of disruption and variations. This technique cannot of course be used in the absence of a routine to calculate calendar dates. It consists simply of the identification of the month in which the activity is programmed to take place and the application of a suitable factor to the stored duration. It is a simple but remarkably effective addition to the traditional analysis, but one which cannot be made without the help of the computer.

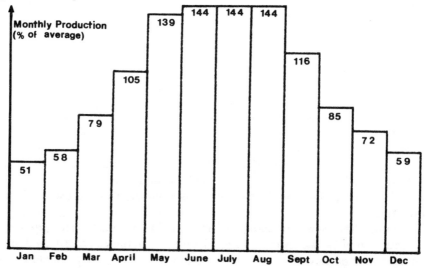

**Figure 5.2** The seasonal variation in construction productivity in the UK.

CONTRIBUTION OF GRAPHICS

## The contribution of graphics to these techniques

The previous chapter showed how graphics techniques can be used to improve the input and output of planning programs and how, in particular, they contribute to the adoption of interactive techniques of computer use. In that chapter the graphics were applied to the basic input and output of planning data. Similar improvements can be brought to the methods described in this chapter.

### Complex links

The dependency matrix was presented in Chapter 4 as a powerful and pleasant way of providing information concerning the structure of the network. The matrix described there, and shown in Figure 4.7, implied the presence only of direct links within the network, but it can be extended to include the complex links described in the early part of this chapter. Two methods are available: to show both the link type and the duration of the link (the lead or lag time) on the matrix itself, or to show only the link type and to allow the user to interrogate the display if he wishes to know further details.

### Project closedown

Project closedown can very easily be shown on the Gantt chart display by shading the vertical column representing the period. The input of closedown data could be done using the same chart, but experience suggests that a more attractive method is to use a calendar block as shown in Chapter 7. The user indicates with his light pen the period of the closedown and this period is shaded on the display.

### Window times

Like project closedown, window times are a time-related phenomenon, but as they are related also to individual activities, the Gantt chart is an ideal medium of data input. A blank Gantt chart is plotted on the graphics screen and the user is invited to indicate any window times he wishes to specify. This he does with a light pen.

### Seasonal variations

If the effects of seasonal variations are to be included, they may well be set to standard values accepted within the organisation, such as were shown in Figure 5.2. If this is the case the values can be stored within the computer program and changed only if special circumstances arise. Such a change can be carried out graphically by displaying the histogram and inviting change. Again change is accomplished through the use of a light pen or a similar device.

## SOME NON-GRAPHICAL PLANNING AIDS

**The pursuit of reality**

The creator of design models is constantly faced with a dilemma: on the one hand he must pursue reality and accuracy so as to make the results of his work acceptable and useful; on the other hand he must guard against making his model so complex that it is difficult or costly to use. The accuracy he obtains must be appropriate to the task in hand; one of the temptations to which the builder and user of a computer model is open is the pursuit of illusory accuracy.

The techniques described in this chapter, most of which are found within existing planning packages, increase realism without enormously increasing the complexity of the program or the time required for it to do its work. Graphical techniques can once again contribute by improving input and output of data and by simplifying the transfer of data between man and machine so that the additional information can be stored without annoying or alienating the user. Thus methods which in the past would not have been used can now help the planner to produce plans which are both realistic and reliable. Chapter 6 shows how these concepts can be built into a working program, and Chapter 8 shows how the graphics are programmed.

# 6
# NETWORK ANALYSIS ON THE COMPUTER

It is to be expected that the elegance of the network model and of its method of analysis should show itself in the algorithms which are required to translate the manual analysis to the computer environment. This expected elegance does, in fact, occur as, it is hoped, will become apparent in the following pages. In this chapter the analysis algorithm for quite complex network models will be developed, listings of part of a program written in Pascal will be presented, the intention being not that they should be lifted entire and complete from the chapter to machine, but that they should provide examples of one way of tackling the analysis of networks.

All the analysis algorithms which are to be presented assume that the activities are presented to them in the correct logical order: that is, that no activity is presented for operation until all its specified precedent activities have themselves been operated upon. As the user cannot (and should not) be expected to test this ordering for himself, it is prudent to provide at the start of any analysis sequence a routine which inspects the activity list and, if necessary, reorders it.

**The sort algorithm**

To sort the data into an order appropriate for analysis the program must produce a list of the new positions of the data elements. This is most easily done by carrying out a series of forward passes through the data list and storing the numbers of the activities which can legitimately be operated upon.

The flow chart (Fig. 6.1) shows the algorithm. The data are read, and as each activity in the data is encountered a check is made to see if it has been successfully stored previously: if it has, then the next activity is considered; if it has not, then a check on its precedent activities is made. If these have all been stored, then the current activity is stored. This procedure is continued, looping

## NETWORK ANALYSIS ON THE COMPUTER

as necessary through the data list until the number of items in the store is equal to the number of activities.

It should be noted that the number of complete loops should never exceed the number of activities. If the number of loops becomes greater than the number of activities, then a logic loop is presented in the network (i.e. $A \to B \to C \to \cdots \to A$) and a warning should be given.

If the array *index*[*nuacts*, 2] is used to record the new order and the precedence data are contained in *prec*[*nuacts*, *nuprecs*], then a suitable algorithm takes the form shown in Figure 6.1 and the following:

```
sign:='i';
k:=1;
j:=1;
repeat
  for h:=1 to nuacts do
  begin
    for i:=1 to 2 do
      index[h,i]:=0;
  end;
  for h:=1 to nuacts do
  begin
    if index[h,1]=0 then
    begin
      flag:='y';
      for i:=1 to nuprecs do
      begin
        if prec[h,i]>0 then
        begin
          if index[prec[h,i],2]=0 then
            flag:='n';
        end;
        if flag='y' then
        begin
          index[j,1]:=h;
          index[j,2]:=j;
          j:=j+1;
        end;
      end;
    end;
  end;
  if j=nuacts+1 then sign:='o';
  if k=nuacts then
  begin
    sign:='o';
    write('WARNING-LOGIC LOOP');
  end;
  k:=k+1;
end;
until sign='o';
```

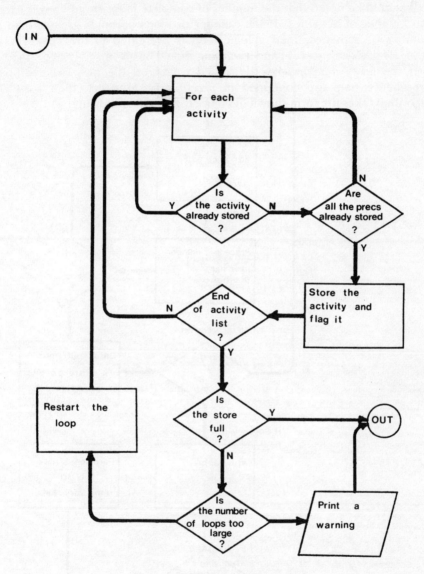

**Figure 6.1** The sort algorithm.

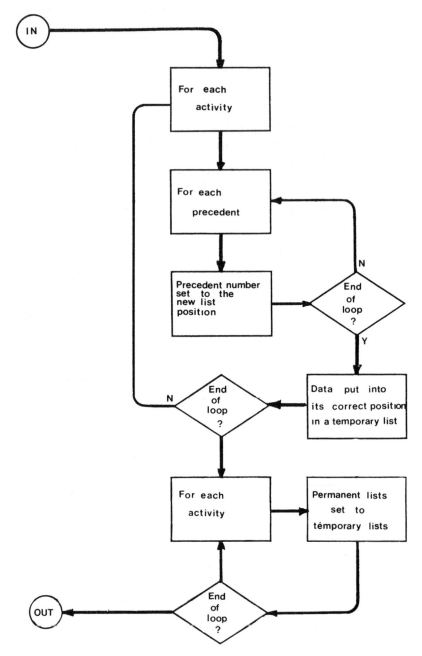

**Figure 6.2** The re-ordering algorithm.

## THE FORWARD PASS FOR SIMPLE NETWORKS

The array *index* now contains both the activity numbers in the required order (in *index*[x,1]) and the list position of the activity numbers (in *index*[x,2]). The programmer can now choose either to reorder his list or to use the index at each call from the data store. The former course, which has the advantage of simplifying the subsequent procedures, will require a routine to convert the various precedent numbers and to restore the data. A flow chart for this is shown in Figure 6.2, and using this the following Pascal algorithm is produced:

```
for h:=1 to nuacts do
begin
  for i:=1 to nuprecs do
  begin
    if prec[h,i]>0 then
      prec[h,i]:=index[prec[h,i],2];
  end;
  for i:=1 to nuprecs do
    tprec[index[h,1],i]:=prec[h,i];
end;
for h:=1 to nuacts do
begin
  for i:=1 to nuprecs do
    prec[h,i]:=tprec[h,i];
end;
```

It should be noted that this algorithm has operated on the list *prec*. In practice other lists built in the same order as *prec* would be changed within the algorithm in the same way that *prec* has been changed.

These two algorithms will take a list of activities and precedents which have been fed into the list in random order, and convert it to a list such that no activity is presented in the list before any of its precedent activities. This is the form which is necessary for the forward and backward pass operations.

## The forward pass for simple networks

The sorting algorithms which have been presented above produce data lists which can be very easily analysed. The forward pass for a simple network (that is one which does not contain complex links) becomes merely a checking against the current value of the early start of an activity the times of the early completion of all its precedents. If the finish of the precedent occurs after the start of the activity then the start of the activity is appropriately altered. This simple procedure is represented by the flow chart (Fig. 6.3) and can be carried out by the following algorithm:

```
finish:=0;
start:=0;
for h:=1 to nuacts do
```

# NETWORK ANALYSIS ON THE COMPUTER

**begin**
  *time[h,1]:=start*
  *time[h,2]:=start+dur[h]*;
  **for** *i:=1* **to** *nuprecs* **do**
  **begin**
    **if** *prec[h,i]>0* **then**
    **begin**
      **if** *time[h,1]<time[prec[h,i],2]* **then**
        *time[h,1]:=time[prec[h,i],2]*;
    **end**;
  **end**;
  **if** *time[h,2]<time[h,1)+dur[h]* **then**
    *time[h,2]:=time[h,1]+dur[h]*;
  **if** *finish<time[h,2]* **then**
    *finish:=time[h,2]*;
**end**;

The finish date of the network is updated throughout the calculation, and clearly is not dependent for its accuracy on the existence of a single finish event.

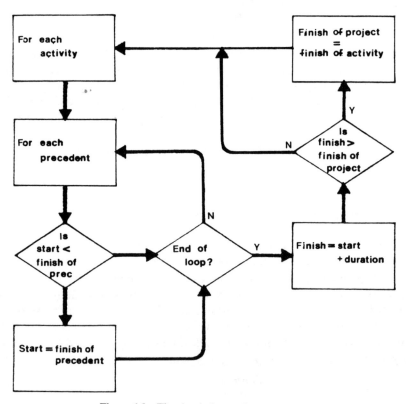

**Figure 6.3** The simple forward pass.

## INCORPORATION OF WINDOW TIMES

Thus multiple starts and finishes can be handled by this algorithm. The backward pass is identical in form to the forward pass: the parameter $h$ is now progressively reduced to produce a backward scanning of the list, and all the event times are set to their theoretical maxima before analysis starts (that is, to the completion date of the project for the finish events and to the completion date minus the activity duration for the start events). The algorithm becomes:

```
for h:=1 to nuacts do
begin
  time(h,3):=finish−dur[h];
  time(h,4):=finish;
end;
for h:=nuacts downto 1 do
begin
  for i:=1 to nuprecs do
  begin
    if prec[h,i]>0 then
    begin
      if time[prec[h,i],4]>time[h,3] then
      begin
        time[prec[h,i],4]:=time[h,3];
        if time[prec[h,i],3]>time[prec[h,i],4]−
                              dur[prec[h,i]] then
          time[prec[h,i],3]:=time[prec[h,i],4]−
                              dur[prec[h,i]];
      end;
    end;
  end;
end;
```

## Incorporation of window times

Window times, discussed in Chapter 5, can easily be included in the algorithms which have just been described. In the algorithms described all the events were set to the earliest possible time at the start of the forward pass, that is to the start of the project, and the analysis then recalculated these items. A similar procedure was adopted for the backward pass. If instead of setting the events to the start and finish of the project, as previously, they are set to the start and finish of the relevant windows, then the analysis will automatically set the events within the required time brackets. This can be done by a simple change to the algorithm:

```
time[h,1]:=start;
time[h,2]:=start+dur[h];
```

is replaced by

```
time[h,1]:=window[h,1];
time[h,2]:=window[h,2];
```

in the forward pass and similarly in the backward pass

$$time[h,3]:=finish-dur[h];$$
$$time[h,4]:=finish;$$

becomes

$$time[h,3]:=window[h,1];$$
$$time[h,4]:=window[h,2];$$

## Complex links

Although complex links are simple in concept, they add considerably to the complication of the analysis algorithm. The data required to specify the linkage must now contain not one element as in the past (the identity of the precedent activity) but three (the identity of the precedent, the type of link, and the period assigned to the link). In practice these three data items can be reduced to two by storing an algebraic value for the period and using the convention that a negative period implies a lag, a positive period a lead. Lead/lag links must be stored as two separate links to the same precedent under this system.

Figure 6.4 is a flow chart for this complex forward pass. The algebraic sign of the period can very conveniently be used to direct the branching at 'type of link'. It should be noted that the lag link affects only the finish event in the forward pass. Using this flow chart the following algorithm can be developed for the forward pass:

```
finish:=0;
start:=0;
for h:=1 to nuacts do
begin
  time[h,1]:=start;
  time[h,2]:=start+dur[h];
  for i:=1 to nuprecs do
  begin
    if prec[h,i]>0 then
    begin
      if comp[h,i]>-1 then
      begin
        if comp[h,i]>0 then
        begin
          if time[h,1]<time[prec[h,i],1]+comp[h,i] then
            time[h,1]:=time[prec[h,i],1]+comp[h,i];
        end;
        if comp[h,i]=0 then
        begin
          if time[h,1]<time[prec[h,i],2] then
            time[h,1]:=time[prec[h,i],2];
        end;
```

## COMPLEX LINKS

```
        if time[h,2]<time[h,1]+dur[h] then
            time[h,2]:=time[h,1]+dur[h];
      end;
      if comp[h,i]<0 then
      begin
        if time[h,2]<time[prec[h,i],2]−comp[h,i]then
            time[h,2]:=time[prec[h,i],2]−comp[h,i];
      end;
      if finish<time[h,2] then
          finish:=time[h,2];
    end;
  end;
end;
```

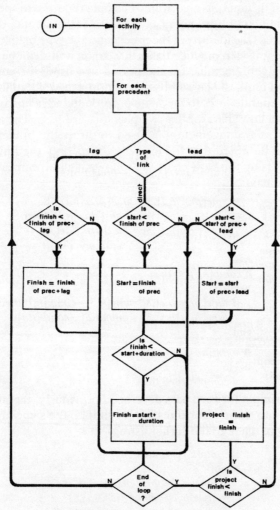

**Figure 6.4** The complex forward pass.

The backward pass again follows the structure of the forward pass but the difference between lead and lag links make significant changes. The flow chart for this algorithm is shown in Figure 6.5. The algorithm for this is:

```
for h:=1 to nuacts do
begin
  time[h,3]:=finish−dur[h];
  time[h,4]:=finish;
end
for h:=nuacts downto 1 do
begin
  for i:=1 to nuprecs do
  begin
    if prec[h,i]>0 then
    begin
      if comp[h,i]<1 then
      begin
        if comp[h,i]<0 then
        begin
          if time[prec[h,i],4]>time[h,4]+comp[h,i] then
            time[prec[h,i],4]:=time[h,4]+comp[h,i];
        end;
        if comp[h,i]=0 then
        begin
          if time[prec[h,i],4]>time[h,3] then
            time[prec[h,i],4]:=time[h,3];
        end;
        if time[prec[h,i],3]>time[prec[h,i],4]
                              −dur[prec[h,i]] then
          time[prec[h,i],3]:=time[prec[h,i],4]
                              −dur[prec[h,i]];
      end;
      if comp[h,i]>0 then
      begin
        if time[prec[h,i],3]>time[h,3]−comp[h,i] then
          time[prec[h,i],3]:=time[h,3]− comp[h,i];
      end;
    end;
  end;
end;
```

As with the simple forward pass developed previously, the incorporation of window times into these algorithms is very simple, the same alterations being necessary as were made to the earlier listings.

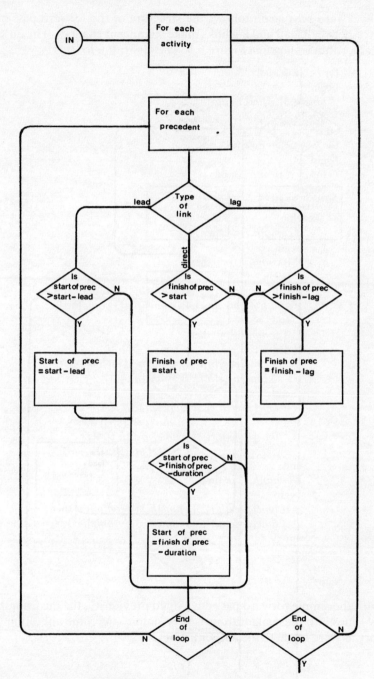

**Figure 6.5** The complex backward pass.

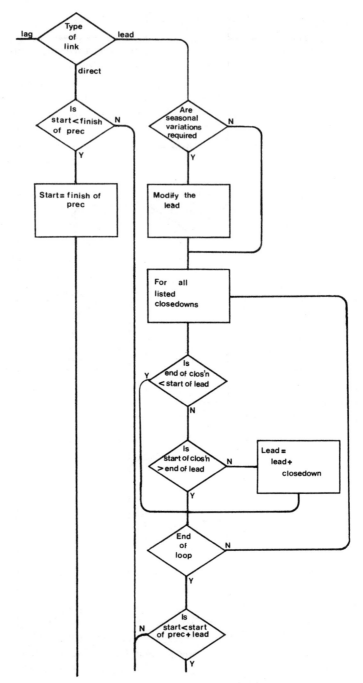

**Figure 6.6** The complex forward pass with closedown and variable production.

THE PART OF ANALYSIS

## Closedown periods and seasonal production variations

Project closedown and seasonal production variation require similar treatment within the analysis algorithms. At each point at which time is being added to a previously computed event time the duration of the time added must be seasonally adjusted and a check must be made to ascertain if the time being added is affected by closedown. If this is to be done the closedown times must be stored in the internal time units, that is in project days rather than in the form day:month:year, and productivity values should also be allotted to bands of project time. The conversion of dates to internal time units will be described later.

If closedowns and productivity are stored in an appropriate fashion, then additions of the form shown by the flowchart (Fig. 6.6) are necessary at each point where time is added (that is at the addition of lead and lag times and at the addition of activity duration). If this extra computation is added to the part of the analysis shown in Figure 6.6, then that part of the analysis algorithm becomes:

```
if comp[h,i]>0 then
begin
  if var='y' then
    period:=comp[h,i]*vary[time[prec[h,i]]];
  for k:=1 to nuhols do
  begin
    if hol[k,1]>0 then
    begin
      if hol[k,2]=>time[prec[h,i],1] then
      begin
        if hol[k,1]<time[prec[h,i]]+period then
          period:=comp[h,i]+hol[k,2]-hol[k,1];
      end;
    end;
  end;
  if time[h,1]<time[prec[h,i],1]+period then
    time[h,1]:=time[prec[h,i],1]+period;
end;
```

## The part of analysis in computer-aided planning

The algorithms which have been presented here are all concerned with the analysis of the network. Vital though these routines are, they will form only a small part of the total computer program, for usually approximately 80% of the program will be concerned with the input and output routines. These other routines will be considered in the next chapter.

# 7
# A PLANNING PROGRAM WITH GRAPHICAL I/O

**The objectives**

The devices and techniques described in the previous chapters have been shown to be of potential benefit to the planner who wishes to use the computer in an interactive way as part of a project method design system. Some of the techniques are available in existing packages, but some, due to their reliance on graphics, are not. This chapter describes a program which is designed to make these techniques available through graphics and to show the usefulness of graphics in programs which are quite simple in concept. The program exploits the graphical capabilities of the new generation of computer hardware which is distinguished from its predecessors by its power, low cost, and high quality and versatile I/O. Graphical techniques are used in this program as a way of ensuring fast and accurate communication between user and machine, a fundamental requirement of interactive design.

Discussions with planners working in industry suggest that planning programs have in the past been less effective than they might have been because of the reluctance of the potential user to enter a field which required an initial expenditure of time. The use of graphically controlled programs lowers this threshold and thus contributes to overcoming this resistance. Experience of menu-driven programs demonstrates how the threshold can be almost swept away by graphics and thus how the effectiveness of planning techniques can be quickly improved.

As hinted above, graphical techniques can benefit the user in two ways: they can make the program easier to run and so take away the need for voluminous manuals, and they can greatly improve the efficiency of data transfer. Both of these are necessary if the computer is successfully to be introduced into the hostile world of construction.

THE OBJECTIVES

## *Program direction*

The use of menus for the control of computer programs is not new. In the past they have contributed greatly to the user-friendliness of non-graphical systems which have been designed to be interactive. In the non-graphical environment they require some keyboard literacy on the part of the operator and although the level of this skill required of the operator may be very low (the typing of single letters or numbers) it may be sufficient to deter those who are not computer enthusiasts. If graphics are available and can remove this deterrent then they should be used.

The indication of choice from a menu by means of graphics can be made in two ways: through the use of menus displayed on the computer screen or by the use of 'paste-on' menus on a digitiser pad. The screen menu can be accessed in several ways: by the use of a cursor controlled by a joystick or button pad, by the use of a light pen, or by the use of a cursor controlled by a digitiser pad or mouse. The joystick-controlled cursor has no advantage over the light pen other than cost; it is slower in operation than the light pen and has the effect of distancing the user from the program. Joystick control is more accurate than light pen control, but this is of no consequence when the cursor is being used for menu selection. Where light pens are not available, however, joystick control is acceptable.

An alternative to light pens is provided by cursors controlled by digitiser pads. In this case the device operates not on the screen itself but on a horizontal board mounted next to the computer keyboard. The digitiser offers two advantages over the light pen: the use of a horizontal surface, which is more natural for most people than the vertical screen of the VDU and which, because it is the resting position of the hand, is quicker than the light pen which has to be picked up, used, and stored away; and the separation of the menu from the screen display allowing the use of 'paste-on' menus. Paste-on menus are being used very successfully by many interactive programs.

In this program the option of the digitiser with on-screen menus was chosen. The choice of the digitiser as against the light pen was based on the ease of use described above and the availability of suitable equipment. On-screen menus were used because of their transportability, the ease of learning which they provide, and the relative simplicity of the menus being used in the program.

The choice of on-screen menus has a considerable effect on the operation of the program. On-screen menus constrain the programmer to menus which are simple and thus lead to a network of menus through which the user proceeds. The disadvantage of this is that it tends to slow the experienced user; the advantage is that the inexperienced user can be gently led through the program by means of the menu network, making the learning process relatively quick and painless. The program direction illustrated here consists therefore of a network of menus through which the user proceeds. By its nature such a network is hierarchical, the needs of the user being defined in progressively increasing detail. The network for the demonstration program is shown in Figure 7.1. Figures 7.2 and 7.3 show the two main menu displays.

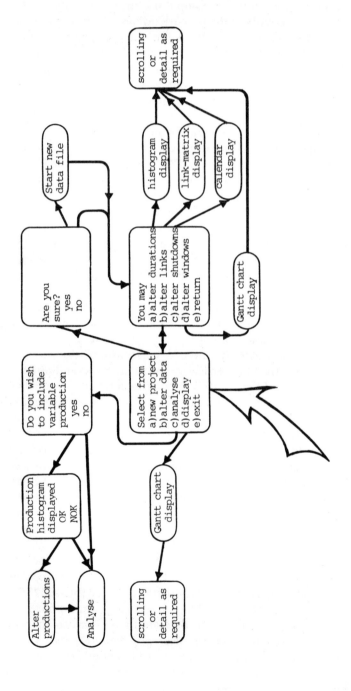

**Figure 7.1** The procedure through the program menu.

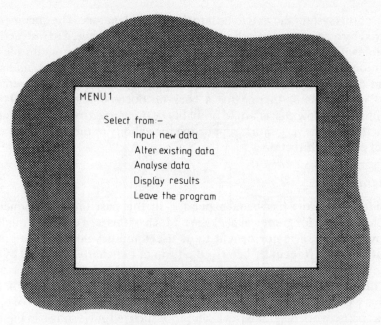

**Figure 7.2** The main menu.

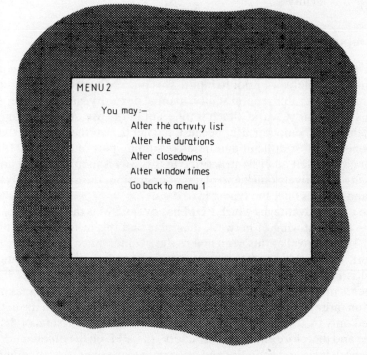

**Figure 7.3** The data input menu.

Use of this system shows it to be largely self-explanatory. The documentation required by a program controlled in this way could be limited to instructions on how to switch on the machine and to gain access to the program plus some indication of the shape of the menu network, such as Figure 7.1, and a 'help' system contained within the program expanding the brief descriptions contained in the menu. In practice a back-up document giving details of the algorithms and flow charts would probably be demanded by the enthusiast. The instruction time is very short, supporting the validity of the decision to use this type of program direction.

*Data transfer*

Considerable effort has been expended in the past on the production of graphical output for graphical packages. Much of this is of excellent quality and is proving to be of major benefit to its users and developers. The graphical output has most frequently been in the form of bar charts produced by means of line printers, and the possible disadvantage of this type of output has already been discussed in this book. As part of the work leading to this book output systems were developed which produced plotted output in the form of both bar charts and network diagrams; an example of such output is shown in Figure 4.4. This output form has been well received by those in industry with whom it has been discussed and it is clear that any commercial package developed should include this facility.

**Data input routines**

In the past the input of data to planning packages has frequently required the drawing of the network prior to input. This process is very time consuming and requires considerable mental agility, manual dexterity, and patience. The need for this preparatory work, which is inherent in the use of arrow diagrams, is a disincentive to computer use, because having drawn the diagram the planner has done the most difficult and time-consuming part of the work. It has been shown earlier that analysis structured in this way is inefficient and unpopular, replacing a relatively quick and pleasant operation, the manual analysis, with a slow and tedious one, the typing in of data.

One of the advantages which PDM has over CPM is the removal of the need for the prior drawing of network diagrams and this in turn brings two major gains. The removal of this need makes the planner much more ready to use his network as a design model: the investment of time in the presentation of the model is much smaller and the planner will be less loath to alter it, and the planner is now able to see the advantage of computer use, again making design iteration more acceptable. In PDM the logical links which form the network are defined only by the activities which they connect; such a system of definition is unique and therefore entirely satisfactory. In CPM, on the other hand, the links define both logic and activities and are specified in terms of event numbers, and the prior numbering of events without the help of a network diagram is difficult.

This advantage of PDM makes it possible to devise an input system which refers only to the activities and their properties and which, because of graphics, is both fast and unambiguous. The data input required by such a system consists therefore only of data concerning the activity itself, data concerning the relationship of the activity with other activities in the network, and data concerning the project environment.

## *Data concerning the activity itself*

The data concerning the activity itself in the program illustrated is limited to details of title and duration. There is no reason why much more information should not be provided including, perhaps, cost, resource load, and materials use, but these items have not been included because the primary purpose is to demonstrate the use of graphics in relatively simple applications. The titles and durations of activities contained in the data files are displayed by means of a horizontal histogram (Fig. 7.4).

The histogram can be scrolled either horizontally or vertically. Durations displayed on the chart are altered by indicating an existing activity; the operator is asked if he wishes to delete the activity from the list or to alter the stored duration. If he chooses the former the activity is deleted; if the latter he is asked to indicate the new value using the cursor, and the diagram is altered to show the change. If the operator indicates a position on the screen lower than that of the last activity in the list, then he will be asked for the title of the new activity and for its duration. The new activity will then be displayed as part of the list.

All instructions to the operator are shown on the screen either above the main display (for instructions which are semi-permanent) or below it (for instructions which refer only to the next action by the operator). The format which has been used for the input of this data has been found to be pleasant to use, fast, and self-checking. Gross errors in duration are extremely easy to identify when the durations are plotted in this histogram form. The changing of data is very fast, especially when a light pen or digitiser-driven cursor is available. Joystick-driven cursors have been tried but have been found to be somewhat frustrating for the user because of their lack of speed. It is felt, however, that even the relatively slow joystick is preferable to the non-graphical data input alternative.

## *Data concerning the relationships between activities*

Although the precedence diagram method allows the construction of network models of project plans without the prior drawing of the network diagram, and so contributes to a major saving in time and effort on the part of the planner, it still requires, of course, that the links between activities be unambiguously specified in some way. In the non-graphical computer environment these links are specified by allotting each activity a unique number and, for each activity, typing into the data files a list of the numbers of the activities upon which the

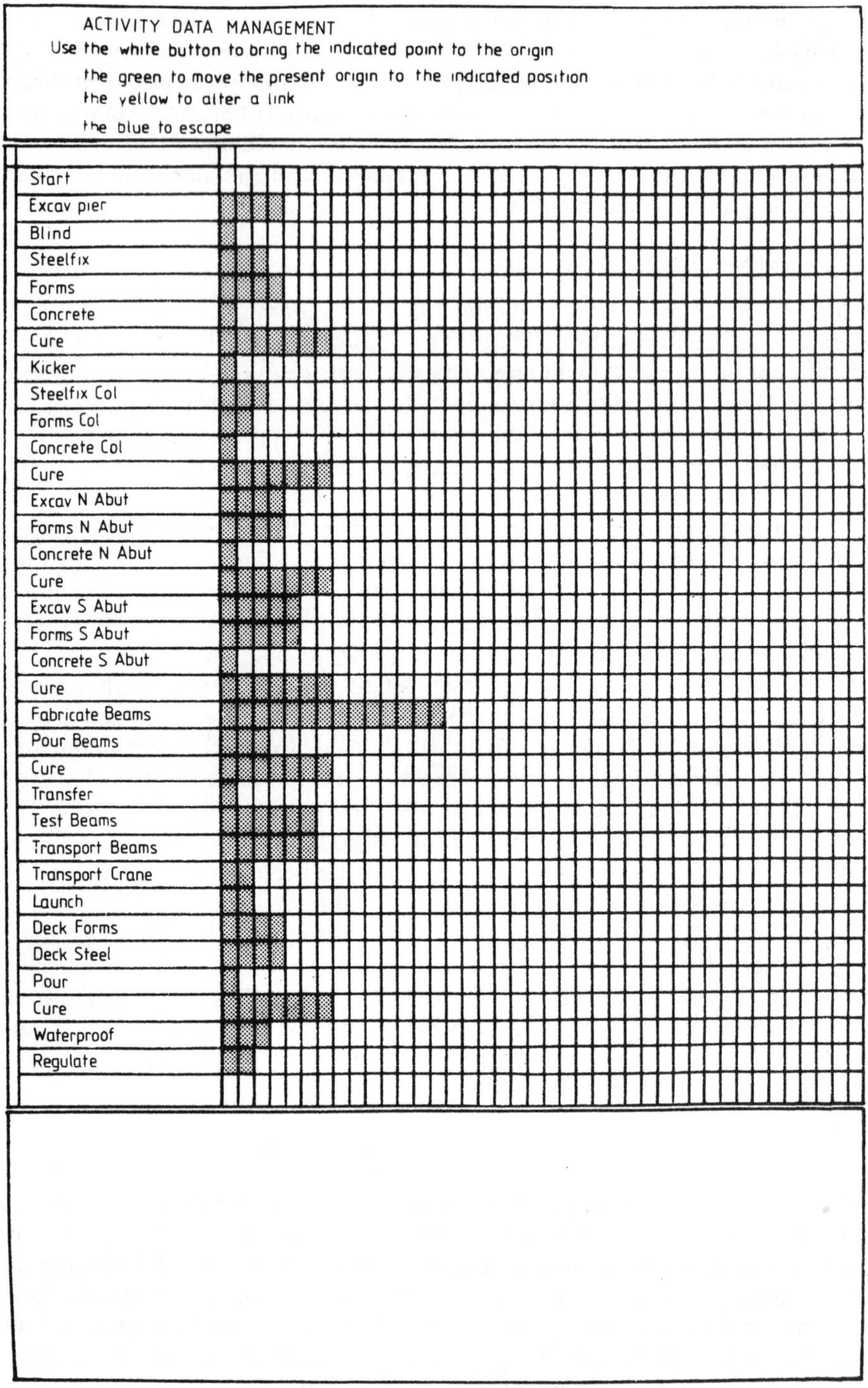

**Figure 7.4** The horizontal histogram.

activity depends. Such a system has been successfully used in many PDM programs.

Non-graphic systems can be pleasant to use if they are well designed but they suffer from the disadvantage of requiring access by the planner to lists of activities in order that he can specify the activity numbers. This access can be achieved in two ways: the planner can prepare a list on a separate sheet of paper as he is inputting the activity details and subsequently use this list as his aide-de-memoire; or the list of activities stored in the data files can be displayed on the screen while the planner is considering the logic of the network. It is obvious that the second of these is the more convenient for the planner and even if only relatively crude terminal screens are available is an input system which should be worked towards.

If graphics are available it is possible to improve on this list display and to remove the need to type numbers at this stage altogether. This different way of showing links is the dependency matrix in which the activities are listed horizontally and vertically along the sides of a grid. Dependency is shown by shading the grid intersections. Dependency grids have been used in the demonstration program and the screen display is shown in Figure 7.5. In the dependency matrix the horizontal axis represents the precedent activities, those upon which the dependent activities, listed on the vertical axis, logically depend. Direct links are shown by a simple block shading, and this block is altered in shape to represent a lead (◢), a lag (◤), or a lead/lag (◧). Figures 7.6–8 show the input of complex links.

It is possible to represent the various types of links by an asterisk for a direct link and a number representing the extent of the lead or lag (negative for a lag) for complex links. This system has the advantage of showing all the information on the screen at the same time, but has the disadvantage of requiring a large grid spacing and therefore much more scrolling than would otherwise be necessary. To avoid the need for frequent scrolling, which is particularly disruptive at this stage of the data input when ideally the planner should have the whole of the dependency matrix in front of him, the lower area of the screen can be used to provide the detail of complex links. Thus if the user requires data concerning a complex link he indicates it with the cursor and a diagram of the relevant part of the network is displayed.

The dependency matrix has been found to be extremely pleasant to use, very complex networks can be built up very quickly and because the speed of alteration is high, some of the disincentive to iterative use is removed. The use of this type of input, in particular, enables the planner to examine the effect of links which are dictated; not by logic, but by managerial expediency, such as the dependence of one activity upon another because of a management decision to limit the availability of a certain resource. In addition to this ease of alteration is the gain due to the possibility, brought by this input device, of the planner thinking at the terminal – the data assembly and the data input phases (see Ch. 2) being brought together so as to remove the frustration of the second and increase the creativity of the first.

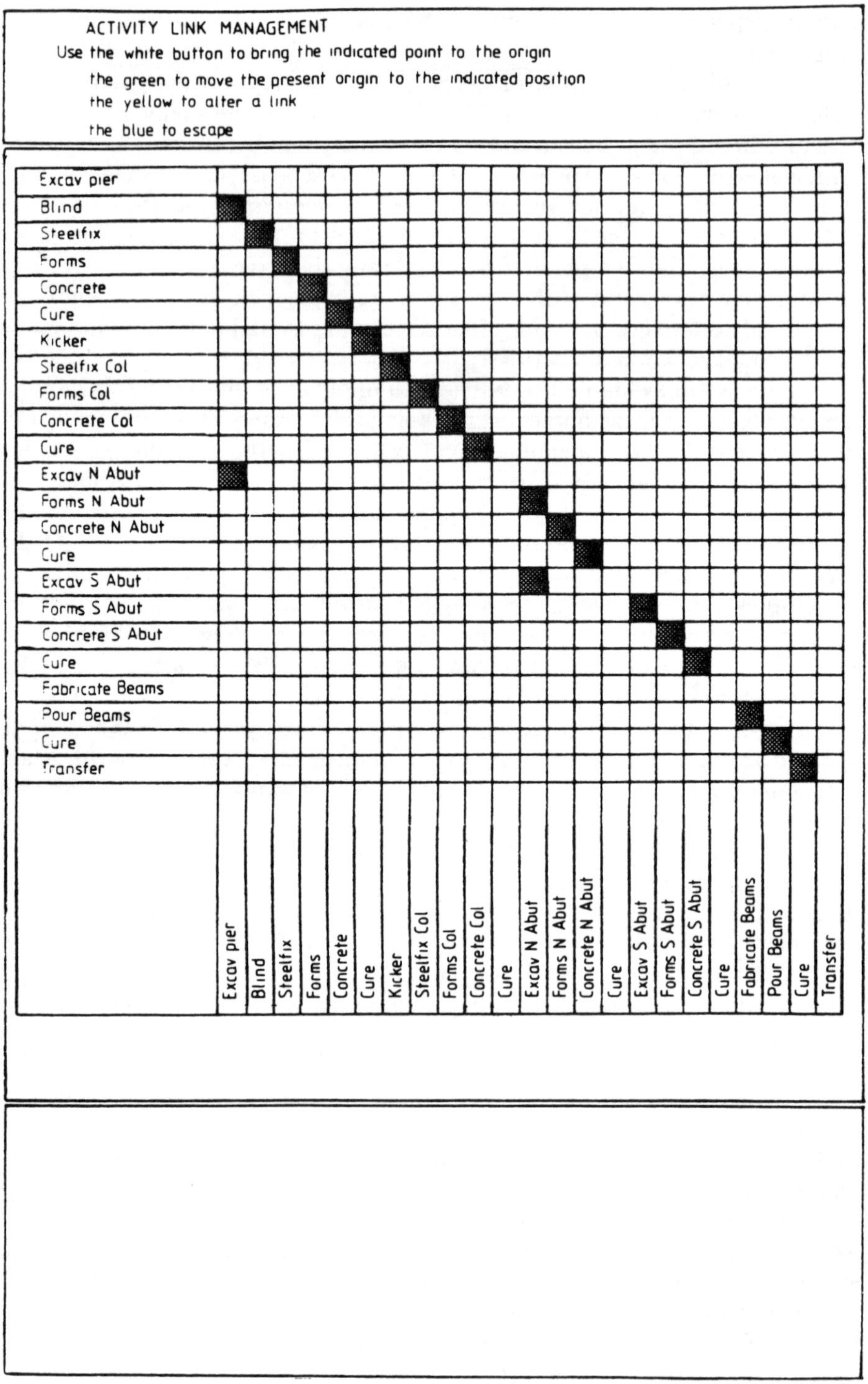

**ACTIVITY LINK MANAGEMENT**

Use the white button to bring the indicated point to the origin
the green to move the present origin to the indicated position
the yellow to alter a link
the blue to escape

| | Excav pier | Blind | Steelfix | Forms | Concrete | Cure | Kicker | Steelfix Col | Forms Col | Concrete Col | Cure | Excav N Abut | Forms N Abut | Concrete N Abut | Cure | Excav S Abut | Forms S Abut | Concrete S Abut | Cure | Fabricate Beams | Pour Beams | Cure | Transfer |
|---|---|---|---|---|---|---|---|---|---|---|---|---|---|---|---|---|---|---|---|---|---|---|---|
| Excav pier | | | | | | | | | | | | | | | | | | | | | | | |
| Blind | ■ | | | | | | | | | | | | | | | | | | | | | | |
| Steelfix | | ■ | | | | | | | | | | | | | | | | | | | | | |
| Forms | | | ■ | | | | | | | | | | | | | | | | | | | | |
| Concrete | | | | ■ | | | | | | | | | | | | | | | | | | | |
| Cure | | | | | ■ | | | | | | | | | | | | | | | | | | |
| Kicker | | | | | | ■ | | | | | | | | | | | | | | | | | |
| Steelfix Col | | | | | | | ■ | | | | | | | | | | | | | | | | |
| Forms Col | | | | | | | | ■ | | | | | | | | | | | | | | | |
| Concrete Col | | | | | | | | | ■ | | | | | | | | | | | | | | |
| Cure | | | | | | | | | | ■ | | | | | | | | | | | | | |
| Excav N Abut | ■ | | | | | | | | | | | | | | | | | | | | | | |
| Forms N Abut | | | | | | | | | | | | | ■ | | | | | | | | | | |
| Concrete N Abut | | | | | | | | | | | | | | ■ | | | | | | | | | |
| Cure | | | | | | | | | | | | | | | ■ | | | | | | | | |
| Excav S Abut | | | | | | | | | | | | | ■ | | | | | | | | | | |
| Forms S Abut | | | | | | | | | | | | | | | | | ■ | | | | | | |
| Concrete S Abut | | | | | | | | | | | | | | | | | | ■ | | | | | |
| Cure | | | | | | | | | | | | | | | | | | | ■ | | | | |
| Fabricate Beams | | | | | | | | | | | | | | | | | | | | | | | |
| Pour Beams | | | | | | | | | | | | | | | | | | | | ■ | | | |
| Cure | | | | | | | | | | | | | | | | | | | | | ■ | | |
| Transfer | | | | | | | | | | | | | | | | | | | | | | ■ | |

**Figure 7.5** The dependency matrix.

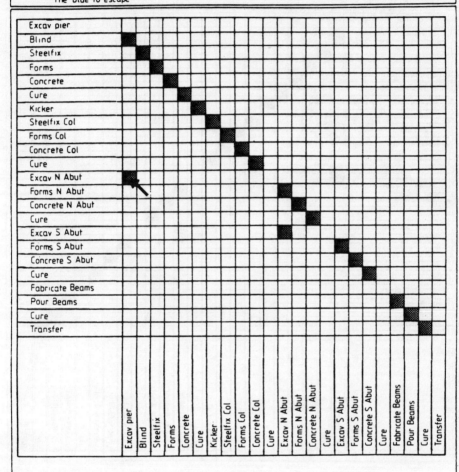

**Figure 7.6** The dependency matrix (continued).

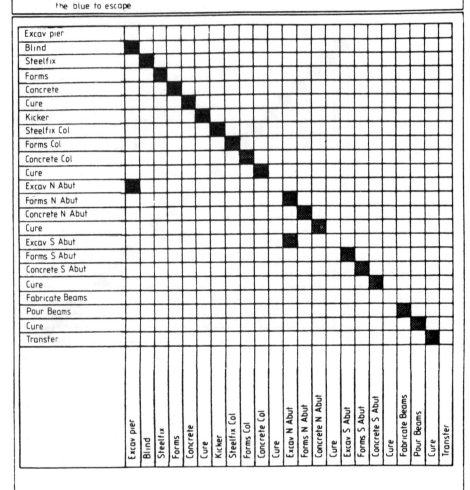

**Figure 7.7** The insertion of complex links.

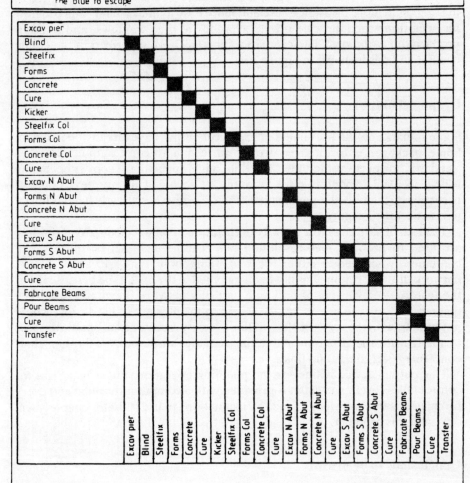

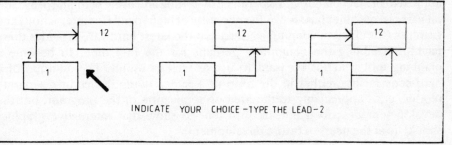

**Figure 7.8** The insertion of complex links (continued).

## A PLANNING PROGRAM WITH GRAPHICAL I/O

### *Data concerning the project environment*

Chapter 5 introduced several devices which could increase the realism of the network model by more accurately modelling the points of contact of the project with the environment; graphics can increase the usefulness of these by making them more pleasant to use and more easy to understand. The program includes three such devices: project closedown, time windows, and seasonal productivity variations. Their incorporation has been described in the earlier chapter and this is shown in Figure 7.9.

### Graphical output

Some remarks on the usefulness of graphical output have already been made. Some packages unfortunately do not have a hard-copy output device as standard; however, at the time of writing, high-quality hard copy is becoming widely available and this must certainly be included in any development, particularly if the planning/control continuum which has already been referred to is to become a reality.

The lack of hard copy forces the concentration upon screen display. Graphics are used to produce high-resolution bar charts which can be quickly and continuously scrolled in either the $x$ or $y$ direction (Figs 7.10 and 7.11). A network drawing routine has not been included in the prototype because it was thought that the viewing of only one part of a network, inevitable when the diagram is displayed on an A4 size screen in the absence of hard copy, would not be helpful to the user. In place of this a facility is provided to enable the user to interrogate the program about the activities displayed on the bar chart concerning precedents and float.

This output device is fine for the planning operation, but it is useless for control unless every site cabin is provided with a computer terminal and every foreman has the knowledge (and the inclination) to use it. Later chapters will deal with this problem.

### Conclusions to be drawn

This program demonstrates the power of both interaction and graphics when applied to project planning. Experience of writing and using planning programs on various machines has shown the necessity of high speed (for interaction) and graphics (for efficient communication) and the latest hardware provides these facilities enabling the computer, perhaps for the first time, to become a planning tool which is pleasant to use as well as useful. The addition of a hard-copy facility, either in the form of a screen-dump device or a separate plotter, adds appreciably to the area of application of the program, but the development described here is sufficient to show that interactive graphics should lie at the heart of future development.

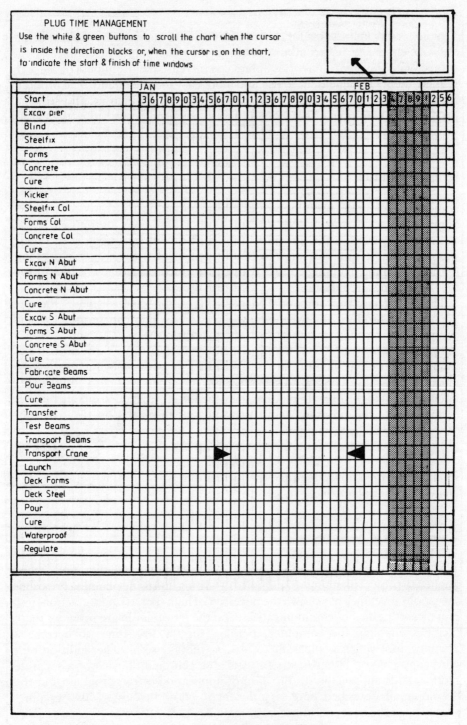

**Figure 7.9** The insertion of 'plug times'.

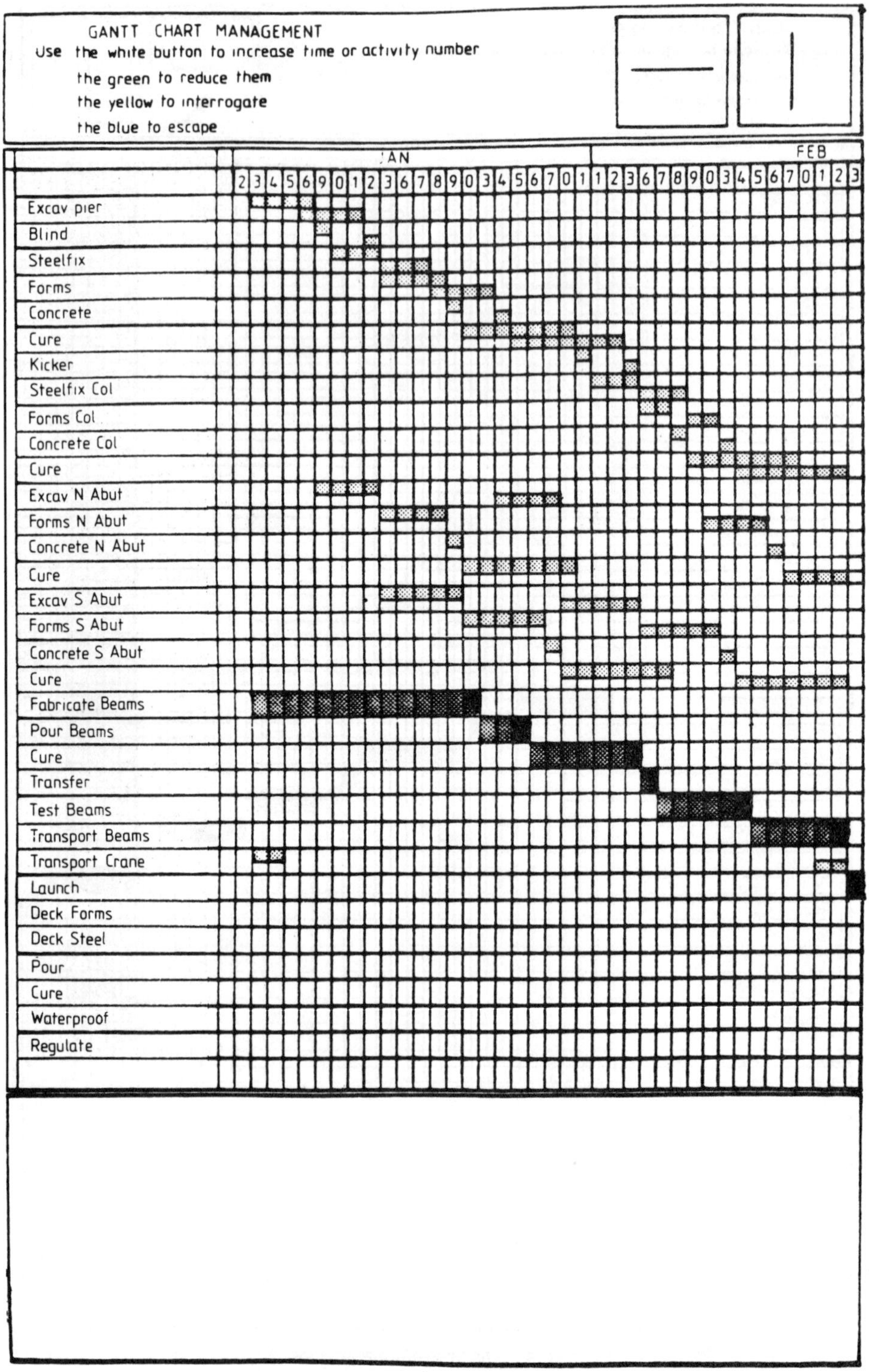


**Figure 7.10** Scrolling the chart.

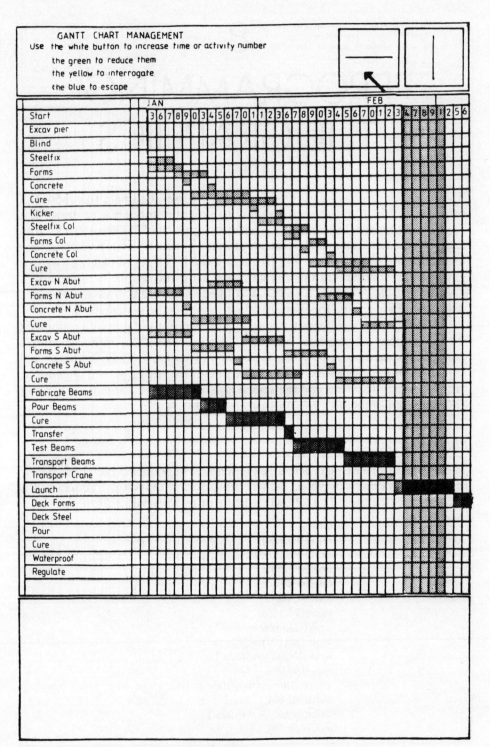

**Figure 7.11** Scrolling the chart (continued).

# 8
# PROGRAMMING THE GRAPHICS

The graphical displays described in Chapter 4 are all simple (composed of orthogonal grids) and as such are easy to program and quick to draw. It is important, however, not to confuse the term 'simple' with the term 'crude'. The key to the production of an interactive graphics system which is popular among users, and therefore widely and frequently used, is a set of carefully written graphics I/O routines. In this chapter I will show the simplicity of the use of graphics for this purpose, producing routines which will draw the required diagrams. They are the foundations of a system, not the system itself.

One difficulty which faces the author of illustrations of graphics routines is the multiplicity of graphics languages. To overcome this difficulty simple graphics instructions specific to this book will be used. They are listed and explained in the Appendix. It should be noted that the on-screen dimensions are stated in millimetres and the screen is assumed to be 200 mm × 140 mm and landscape-oriented.

**Menu direction**

The setting up of a menu is of course the simplest of graphical exercises, the list of available options is drawn on the screen together with such instructions as are necessary for the user. The following simple program would achieve this:

> *clear;*
> *move(60,120);*
> *line(140,120);*
> *line(140,20);*
> *line(60,20);*
> *line(60,120);*
> *move(70,110);*
> *write('Do you wish to:');*
> *move(80,100);*
> *write('Input new data');*
> *move(80,90);*
> *write('Alter existing data');*
> *move(80,80);*
> *write('Analyse the data');*
> *move(80,70);*

## MENU DIRECTION

> write('Display results');
> move(80,60);
> write('Leave the program');
> move(70,50);
> write('Indicate your choice using');
> move(70,40);
> write('the light pen');

This display is shown in Figure 8.1.

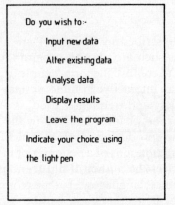

**Figure 8.1** The menu display as programmed.

It is now necessary to read the intentions of the user, this is done by:

> **repeat**
>     flag:='i';
>     coord(i,j);
>     k:=(j−100) **div** 10;
>     **if** k>0 **then**
>     **begin**
>         **if** k<6 **then**
>         **begin**
>             **case** k **of**
>                 1:input;
>                 2:alter;
>                 3:analyse;
>                 4:display;
>                 5:exit;
>             **end**;
>             flag:='o';
>         **end**;
>     **end**;
> **until** flag:='o';

Clearly this routine can be used to lead to another menu display and thus produce the hierarchy of menus which is necessary for computer programs.

## Graphical input

### The input of activity parameters

The 'histogram' input form which was described in Chapter 4 is pleasant to use. It is also simple to program, as follows:

```
clear;
for i:=1 to 25 do
begin
   move(55+i*5,120);
   line(55+i*5,29);
end;
move(20,120);
line(20,29);
for i:=1 to 14 do
begin
   move(20,127−i*7);
   line(900,127*i−7);
   move(22 122−i*7);
   for j=1 to 20 do
      write(name [i,j]);
      block(60 127−147,60+time[i,1]*5, 127−i*7);
end;
move(20,25);
line(180,250);
line(180,5);
line(20,5);
line(20,25)
move(22,20);
write('Alter or add to the display by indicating a new duration');
move(22,15);
write('for an existing activity or a blank line respectively');
move(22,10);
write('exit by pointing to the direction block')
```

This routine will produce the display shown in Figure 8.2.

The action of the user with the light pen will govern what follows. Three options are available: to alter an existing duration, to add an activity to the list, or to leave the routine. A simple routine is:

```
repeat
   coord(i,h);
   if h>25 then
   begin
      k:=(134−h) div 7;
      if k<=nuacts then
      begin
         eblock(60,127−k*7,60+time[k,1]*5,120−k*7);
         time[k,1]:=(i−60) div 5;
         block(60,127−k*7,60+time[k,1]*5,120−k*7);
      end;
```

## GRAPHICAL INPUT

```
    if k>nuacts then
    begin
      move(22,20);
      write('Type in the title of the new activity        ');
      move(22,15);
      write('ending with a full stop                      ');
      move(22,10);
      write('                                              ');
      i:=1;
      repeat
        read(a);
        if a<>'.' then
        begin
          name[nuacts,i]:=a;
          move(20+i*2,122-nuacts*7);
          write(a);
        end;
        i:=i+1;
      until a='.';
      move(22,20);
      write('Indicate the duration of activity            ');
      move(22,15);
      write('                                              ');
      for i:=1 to 20 do
        write(name[nuacts,i]);
      coord(i,j);
      time[nuacts,1]:=(i-60) div 5;
      block(60,127-k*7,60+time[k,1]*5,120-k*7);
    end;
  end;
until h<25;
```

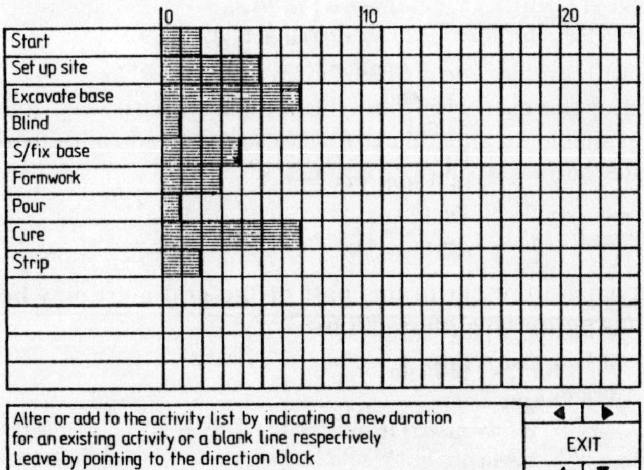

**Figure 8.2** The activity input display as programmed.

## PROGRAMMING THE GRAPHICS

Clearly this routine must be extended so as to accommodate more than fourteen activities, and to do this a scrolling routine must be incorporated. This is achieved by providing a small scrolling menu, as shown in Figure 8.2 and by redrawing the screen, on cue from the menu, for values of $i$ from 10 to 24, 20 to 34, etc.

## *Input of activity precedence*

The advantages of graphical input of the structure of the network have been discussed in Chapter 4. The input device, the dependency grid display together with the necessary hardware (either light pen or remotely controlled cursor) is easy to use and, because of the simplicity of the basic diagram, easy to program. The first stage is to set up the grid representing the top left-hand corner of the complete matrix and to annotate it.

```
clear;
for i:=1 to 14 do
begin
  move(20,142−i*7);
  line(151,142−i*7);
  move(53+i*7,135);
  line(53+i*7,4);
  if i=14 then
  begin
    if i<=nuacts
    begin
      move(22,144−i*7);
      for j:=1 to 20 do
        write(name[i,j]);
      setvert;
      move(51+i*7,6);
      for j:=1 to 20 do
        write(name[i,j]);
      sethor;
    end;
  end;
end;
move(20,135);
line(20,4);
line(151,4);
```

The dependencies present in this part of the grid must now be shown by blocking in the appropriate intersections:

```
for i:=1 to 14 do
begin
  if i<=nuacts then
  begin
    for j:=1 to nuprecs do
```

## GRAPHICAL INPUT

```
begin
   if prec[i,j]>0 then
   begin
      if prec[i,j]<14 then
         block (53+prec[i,j]*7,142−i*7
                60+prec[i,j]*7,135−i*7);
      end;
   end;
 end;
end;
```

Finally, an instruction window such as that shown in Figure 8.3 is drawn on the screen.

The user of this screen has several options open to him. By indicating a vacant intersection space he can establish a new link in the network, by indicating an intersection which is already blocked he can alter an existing link and by indicating the menu in the instruction window he can scroll the screen up or down, backwards or forwards or can leave this part of the program.

Clearly, the first information which must be processed is the intention of the user. This is done as follows:

$coord(i,j)$;
if <151 then $glag:='o'$;
if >150 then
begin
   $flag:='e'$;
   if $j>32$ then

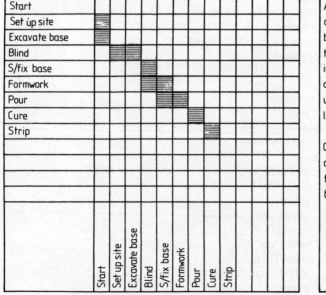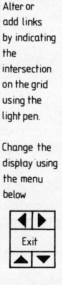

**Figure 8.3** The dependency matrix as programmed.

## PROGRAMMING THE GRAPHICS

```
      begin
        if i<175 then flag:='b';
        if i>174 then flag:='f';
      end;
      if j<22 then
      begin
        if <175 then flag:'u';
        if >174 then flag:='d';
      end;
    end;
```

The value of the variable *flag* now being used to show whether *o*peration, *b*ackwards, *f*orwards, *u*pwards, or *d*ownwards movement is required or if it is necessary to *e*xit. The scrolling is easily done by redrawing the screen using different ranges of activity numbers. The alteration of the screen data is done by the routine:

```
if flag='o' then
begin
  sign:='n';
  h:=(142-i) div 7;
  k:=(j-53) div 7;
  for n:=1 to nuprecs do
  begin
    if prec[k,n]<=h then sign:='y';
  end;
  if sign='n' then
  begin
    for n:=1 to nuprecs do
    begin
      if prec[k,n]>=0 then
      begin
        prec[k,n]:=h;
        n:=nuprec
      end;
    end;
  end;
  if sign='y' then
  begin
    move(25,35);
    write('Do you wish to');
    move(25,28):
    write('delete this link');
    move(25,21);
    write('yes    /    no');
    coord(i,j);
    if i<40 then
    begin
      n:=0;
```

## GRAPHICAL OUTPUT

```
            repeat
              n:=n+1;
            until prec[k,n]=h;
            eblock(53+h*7,142−k*7,46+h*7,135−k*7);
            prec[k,n]:=0;
          end;
          if i>40 then
          begin
            move(25,28);
            write('input special links');
            coord(i,j);
          end;
          if i<40 then
          begin
            inputleadlag
          end;
        end;
      end;
```

The input of special links (through the procedure input lead lag) requires a separate window space. The instruction window or the area between the activity lists can be used for this purpose. Assuming the latter the following routine would be appropriate:

```
              eblock(21,43,59,6);
              move(25,35);
              write('Do you wish to');
              move(25,28);
              write('insert a:−');
              move(30,21);
              write('lead');
              move(30,14);
              write('lag');
              move(30,7);
              write('lead/lag');coord(i,j);
              k:=j div 7;
              case k of:
                1:lead;
                2:lag;
                3:lead/lag;
              end;
```

Each of these subroutines can then be used to receive a value of the time parameter to store the data, and to change the display so as to indicate the special link.

## Graphical output

The basic output form used in the construction industry is the Gantt chart. Such charts have been manually produced and used in the construction industry for

# PROGRAMMING THE GRAPHICS

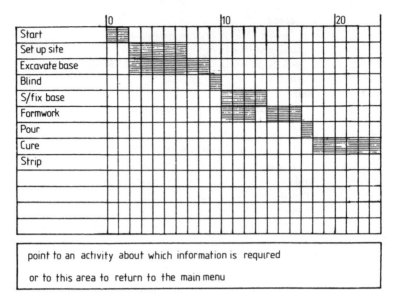

**Figure 8.4** The Gantt chart display.

very many years. They are well understood and widely accepted. In addition to Gantt charts it is useful to produce network diagrams. This latter facility is particularly important where, as in the case of the programs being developed in this book, the planner is not obliged to draw a network prior to the input of data.

The production of these two charts presents no major difficulty although the network display has been found to be effective only where hard copy facilities are available (enabling the whole of the chart to be examined at one time).

The Gantt chart, like many other charts, is inherently simple. A routine for drawing the top left-hand corner of the complete chart has the form:

> *clear*;
> **for** $i:=1$ **to** 25 **do**
> **begin**
>   *move*(55+$i$*5,120);
>   *line*(55+$i$*5,29);
> **end**;
> *move*(20,120);
> *line*(20,29);
> **for** $i:=1$ **to** 14 **do**
> **begin**
>   *move*(20,127−$i$*7);
>   *line*(900,127−$i$*7);
> **end**;
> *move*(22,122−$i$*7);
> **for** $j:=1$ **to** 20 **do**
>   *write*(*name*[$i,j$]);
> **for** $k:=1$ **to** 2 **do**

## GRAPHICAL OUTPUT

```
begin
  if time[i,k*2]<25 then a:=time[i,k*2];
  if time[i,k*2]>25 then a:=25;
  if time[i,k*2+1]<25 then a:=time[i,k*2+1];
  if time[i,k*2+1]>25 then a:=25;
  block(60+a*5,127-i*7-(k-1)*4
        60+b*5,124-i*7-(k-1)*4);
end;
```

This routine, almost identical to that used for the activity parameter input, will produce a chart such as shown in Figure 8.4. The chart should be scrolled in the same way as has been described for the other displays.

As the Gantt chart does not fully represent the network model (it does not, for example, show the logical structure) it is helpful to provide the user with a means of interrogating the display. This is done using the light pen, the user being invited to indicate an activity about which more information is required:

```
repeat
  flag:='y';
  move(20,25);
  line(180,25);
  line(180,5);
  line(20,5);
  line(20,25);
  move(22,20);
  write('point to an activity about which information is required');
  move(22,15);
  write('or to this area to return to the main menu');
  coord(i,j);
  if j<25 then flag:='n';
  if j>25 then
  begin
    k:=(134-j) div 7;
    if k<nuacts then
    begin
      move(22,20);
      write('Activity name-');
      move(92,20);
      for i:=1 to 20 do
        write (name[k,i]);
      for h:=1 to nuprecs do
      begin
        if prec[k,j]>0 then
        begin
          move(52,15);
          write('precedents');
          for i:=1 to 20 do
            write(name[prec[k,h],i]);
      end;
```

>            **end;**
>         **end;**
>      **end;**
> **until** *flag*='*n*';

## The programming of graphics

The purpose of this chapter has not been to describe in detail a graphics-based computer program but rather to demonstrate the ease with which the graphics can be programmed. Project planning is particularly suitable for this approach because of the simplicity of its basic chart. It must be reiterated, however, that simplicity and crudity are not synonymous and great care must be taken in the production of programs to ensure that the programs are easy to use, and above all, foolproof.

# 9
# THE UNCERTAINTY OF CONSTRUCTION PLANNING

**The sources of uncertainty**

Asimow's definition of design as a process which takes place in extreme uncertainty is particularly applicable to construction management. The construction project is unique; the planner of such a project is faced with problems which have never before occurred in exactly the form in which they are presented to him, and thus he is obliged to make forecasts of actions and performance which rely on data to only part of which he has access. It has been pointed out (in Ch. 1) that these data will become available to him as the work progresses and that this continuous flow of data is a prime reason for regarding the planning and control functions of management as two phases of the same operation. The continual flow of data brings the possibility of a continual addition of detail and results in project plans which are dynamic and have information contents in the form of the information wedge shown in Figure 1.2.

The planner cannot, in the face of the uncertainty of the data available to him, simply abandon decision making. The detailed data on which he is dependent is generated only when the action is begun, and the action will not start if all decisions are postponed until all is clear and certain. The planner must at least formulate a strategy for the project. In most cases he must do significantly more than this and provide, for example, cost estimates of accuracy sufficient for the purposes of tendering. The planner is faced with the worrying task of basing important decisions on what is at best incomplete data.

**Increasing the accuracy of data**

In some circumstances the planner of construction projects can increase the confidence he has in the data by initiating further study of the problem. He could, for example, commission extra site investigation studies or obtain more detailed information about the supplies of materials or labour in the area of the site. In general, however, this gathering of extra data may be difficult to justify, for it is expensive and time consuming. Investment in data gathering is subject to severely diminishing returns because some uncertainties are inherent in the construction activity itself and as accuracy in other data areas is increased these inherent uncertainties begin to swamp the estimate.

The cost of pre-project data gathering is an important constraint for the planner for, under the competitive tendering system which is normal in the UK, there is less than a one in six chance that a given tender will be successful. For five tenders out of six the planning work is abortive and the money spent wasted. The extra costs will be covered by increasing the overheads included in the next tender. If pre-project planning costs are too high overheads will increase with respect to those of competitors and the chances of success will fall, increasing the problem. The planner must therefore aim to optimise the accuracy/cost equation and be prepared to work with data which are less than certain.

In addition to the cost constraint on pre-project planning there is a severe time constraint. The competitive tendering system allows very little time for the gathering of additional data. For few projects would the planner be allowed more than twelve weeks to familiarise himself with the project and the area in which the site is located. In these circumstances the planner must work with little more information than that provided by the client and contained in the contract documents and that specialised knowledge of his craft that he has gained through experience of similar projects.

Obviously the situation concerning the gathering of data is less difficult if the project is being planned for an environment different from that of the construction industry competitive tender. Time and cost constraints are less severe in these circumstances but they are still there, appearing now as the problem, familiar in other design disciplines, of knowing when to stop thinking and start doing. Even in this favourable environment detailed deterministic data are not available concerning every parameter for, as has been remarked, some uncertainties are an inherent part of the activity of construction.

## Inherent uncertainties

Although the quality of most data can be increased by the gathering of more detailed information, the value of such an exercise is limited by the inherent and irreducible uncertainty of some of the project parameters. The presence of one uncertain component in a sum prevents the elimination of uncertainty from the total and reduces the data enrichment value of other components. In these circumstances there is a level of accuracy which it is futile to try to exceed, this has been cogently argued with special reference to cost estimation by Lichtenburg (1974). In the construction industry, therefore, it is possible to increase the accuracy of some of the data available to the planner, but not the accuracy of all the data, and these inherent uncertainties, as Lichtenburg showed, affect the whole policy of pre-project planning and cost estimation.

The most obvious source of inherent uncertainty in this country is the weather. This is difficult to predict in the short term, impossible in the long term, and although the planner can reflect broad expectations in his plans (one method of which has been shown in Ch. 5) the unpredictability of detail must somehow be reflected in his plans. Similarly the effects of erratic materials supply, labour unrest, and plant breakdown must be included in the plans, even though the existence or extent of these events cannot be known.

In view of the presence of uncertainty and the difficulty of its removal, the planner must be aware of its effects. He may, knowing the effects, choose to ignore them, but this should be a decision consciously made and not one made by default.

## Handling uncertainty – present practice

The similarities between project planning, the design of project methods, and structural design have already been pointed out in Chapter 2 of this book. There they were used to show the type of design tool which is required by the project planner; they can also be used to show the ways in which uncertainty can be handled. The structural engineer will build an allowance for uncertainty into his design by overdesigning to avoid catastrophe. He spends money to avoid risk, weighing the certain expenditure of a comparatively small amount against the possible expenditure of what could be a very large sum indeed.

The amount of risk which the designer and his client will be prepared to take will be dependent upon the cost of failure and the cost of prevention. The cost of prevention must be such that it is less than the expected cost of failure (that is the cost of failure multiplied by its likelihood). Although it is convenient to express this 'economic' level of risk in these simple terms, the evaluation of the acceptable level is very complex if not, in any but the most theoretical terms, impossible. The cost of catastrophe is very difficult to calculate, for it includes not only the monetary cost of replacement of the lost asset, but also the cost of human suffering, the cost of lost time, and, most ephemeral of all, of lost pride. Normally the structural engineer does not need to carry out this difficult calculation; the allowance for risk will be included in the design codes which he uses, and he can proceed in the (perhaps false) security that the codes give him.

Two approaches to the handling of uncertainty in structural design were described in Chapter 2. Both of these approaches result in the structural designer overdesigning to allow for risk, but they approach the problem from different sides. The allowable stress method, which has been the accepted design method for the whole of the history of structural engineering until very recent times, concentrates on the loads and stresses which are expected to occur. It is assumed that if these loads and stresses are sufficiently small, then collapse will not occur.

The alternative method, the limit state method, concentrates on the event which is to be avoided and has the designer work towards a state of collapse at a given load which is higher than the expected load by a specified amount. This second method has the advantage of dealing with the varying parameters directly and thus makes the assessment of risk and the expected value calculation described above much easier to carry out.

There are no design standards for project method design, the planner must make his own assessment of risk and adjust his plan accordingly. Usually his task will appear to be simpler than that of his structural colleague; it would be unusual indeed for human life to be one of the factors to be included in the calculation and the planner will be working against the twin objectives of cost and time. (Of course the cost of human life may well be included implicitly when unsafe practices are included in project plans, and there is clearly scope for some detailed work here.)

Extra resources reduce time but increase cost; reduced time means reduced risk of overrun with the associated cost to the client. Thus, just as in the structural engineering case, the project planner is faced with the balancing of a certain cost against an uncertain but much larger cost; the shape of the problem is identical in the two cases.

It seems to be very unusual for the planner of construction projects to apply a limit state approach to the uncertainties with which he must work. It is likely that he will fix the finish date of the project, or some other criterion of success, so as to give himself the float which he thinks he will need, and that he will then carry out his planning ignoring the uncertain nature of the constituent elements completely. This approach is directly analogous to the allowable stress method of structural design and shares with it the disadvantage that the assessment and evaluation of risk is very difficult.

It may be argued with some justification that the allowable stress method just described is sufficiently accurate for what is at best a very inexact science. This may be the case, but if it is, then the more sophisticated planners must be careful about using some of the advanced optimisation techniques which are increasingly becoming available. The use of these techniques with deterministic data may be found to be analogous to the building of grand palaces on foundations of sand.

## Handling uncertainty in planning – alternative approaches

The abandonment of allowable stress design procedures because of the difficulty of risk evaluation has led the structural design community towards limit state design as its standard. This type of design allots factors to the various design parameters which reflect the certainty of the parametric value and thus enable a design to be produced with a known risk of failure. This approach as it is laid down in the various codes of practice, British Standards or similar is, of course, an approximation to the actual risk analysis for the structure which, involving many more risk elements than would normally be considered by the user of design codes, would be extremely complex. It does, however, enable the designer quickly to predict the worst case which he can conceive. Its disadvantage is that by stressing the pessimistic it ignores the likely, and it is the likely which, more for the planner than the structural engineer, exercises the mind of the designer.

In these circumstances, where the planner is most concerned with the likely but must be aware of the pessimistic, there seems to be no alternative to a method which carries the uncertainty of the data elements through the analysis and presents an uncertain result for consideration and evaluation. If this type of approach is to be followed, then decisions must be made concerning the detail of uncertainty which is to be carried through the analysis rather than dealt with as a separate analysis of risk.

## Levels of uncertainty

The inherent uncertainties of construction management can affect the plan in two ways: they can affect the productivity of the operation and thus the duration of the activity, or they can affect the activities which the manager chooses to use, thus the strategy of the plan itself. In theory there is no reason why both of these cannot be incorporated into a single network analysis, the first by the specification distributions of productivity and therefore distributions of duration (as used in PERT), the second by the use of decision nodes within a network (that is network nodes which act as branching points). Such nodes are familiar from their development within decision tree systems and are available for example within program GERT. If both techniques are used within a program the project model will become very complex and as such difficult to use, thus the approaches should be fully appraised before being included in a planning package.

### *Specified distributions of productivity*

When a planner sets up his network he must estimate durations and resource levels for the various activities. To do this he will use data from various sources. None of these data constitute an exact and certain estimate of what will happen in the project itself. The planner can overcome this difficulty either by ignoring it at the analysis stage and carrying forward a mental note that the figures in the plan are estimates only and that some variation is likely, or by including the range of possible data values into the analysis. The consideration of the various design philosophies above suggests that this second approach has much to commend it.

In setting a range for productivity the planner is usually implying the presence, in a lower level of the planning hierarchy, of a decision node.

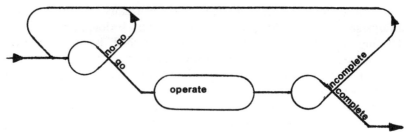

**Figure 9.1** A network representing a go/no go decision.

Consider, as an example of this, the estimation of the productivity of an earthworks operation.

In civil engineering in the UK in the summer it is usual to include a 16% loss of time due to the effects of inclement weather. Thus an operation which would be completed in 24 days of fine weather is allotted a duration of 28 days for the purpose of planning calculations. In making this adjustment the planner allows for the inherent uncertainty of earthworks productivity in our climate.

The planner knows that the operation cannot be completed in less than 24 days and that the most likely completion time is 28 days, but he does not know what a realistic maximum time should be, because in theory the operation could have an infinite duration. A detailed look at the assumptions the planner has made can help to clarify the situation.

The earthworks activity can be represented by the network shown in Figure 9.1. In order to complete the earthworks one day's operations must be repeated 23 times: the project activity cycles continuously through the network and at each cycle, that is at the start of each day, the decision node is used to decide, on the basis of the weather, whether or not earthworks activity is possible. If so, then activity takes place; otherwise the network loops back awaiting the start of a new day.

This network shows that in order to estimate durations the planner must calculate the likelihood of $x$ wet days in a period equal to $24 + x$. Thus the reduction of the problem to a series of go/no go decisions enables the methods of classical statistics to be used. For this particular problem it can be shown that the probability of the duration of the operation being $d$ is

$$P_r^{(d-j)}(1-p_r)^j \frac{d!}{j!(d-j)!}$$

(where $P_r$ = probability of rain and $j$ = the required job length).

This formula can be used to construct a distribution for the duration of the activity and in this case results in the distribution shown in Figure 9.2.

In this simple case the methods of classical statistics could be used once the problem had been modelled using a decision node. This is not always possible and in such cases it may be helpful to use a simulation method to produce the distribution to be used in the project network.

## LEVELS OF UNCERTAINTY

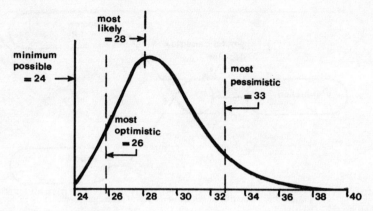

**Figure 9.2**  The distribution of completion time for a 24 period go/no go operation.

The duration distribution is a convenient way to carry the information produced by the implicit decision nodes back to the main network. The carrying back could be done either outside the computer, by carrying out a separate calculation as has just been done for the earthworks example, or within the computer system by means of a hierarchically organised program, but in either case the information carried back should be enough to define the distribution. Thus in the well-known PERT program the range of the distribution and its form is transmitted by the specification of most optimistic, most pessimistic and most likely times, and the distribution is assumed to be of the beta form.

## *The use of decision nodes*

The example quoted above showed how decision nodes could be used within a simple network which was set up to provide information for the main network analysis. An alternative to this approach is to use decision nodes within the network itself.

Usually all the activities in a network are carried out once and once only. It is possible, however, as was shown in the example, to construct networks in which activities may not take place at all or in which activities are repeated. In these networks points are reached where a choice is to be made and the manager, acting on the data available to him at the time, dictates the path which the project will follow. In theory all the decisions which the manager has to make during the course of a project could be included in a network in this way, the network becoming an amorphous mass of interconnecting networks, but it is clear that this approach would not result in the simplicity required of a usable model. As has been noted previously, the planner must make decisions even though the data are uncertain, and the inclusion of multiple alternative paths in the network at strategic level will not help him to do this.

This doubt about the practicality or wisdom of trying to introduce all possible decisions into a network model need not imply that the use of decision nodes in

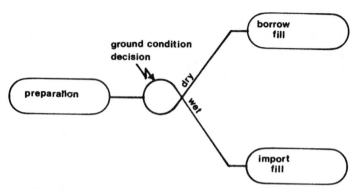

Figure 9.3  The use of decision nodes at strategic level.

the top level of a network is always foolishness. It may be that at the planning stage of a project the planner simply does not have the data required to enable him to recommend one strategy as against another. In these circumstances he can either draw up alternative plans or he can include both possibilities in one network by the use of a decision node.

For example it may be that the earthworks activity referred to previously can be carried out in two ways, by borrowing fill from outside the site and tipping the cut from the site, or by carrying out a cut and fill operation. The decision may rest on the condition of the excavated material, which the planner cannot know until the operation starts. In this case the inclusion of a decision node in the network may be helpful; Figure 9.3 shows how this would be represented in the network.

This use of decision nodes to represent alternative strategies within one network gives a more realistic representation of the project than would be obtained by, for example, representing the earthworks as an unspecified activity of duration equal to the weighted average of the alternative durations, and the results of the analysis of such a representation will be easier to use. The decision on the number of such nodes in a network depends on the planner's preference and, perhaps, his ability to handle complexity. Intuition suggests that unless the number is set very low incomprehensible networks will be produced. It should be noted, of course, that this difficulty is not present in a hierarchically structured network, and it may be that only by combining hierarchies and decision nodes will a satisfactory method be found of handling this strategic uncertainty.

Programs are available which will handle decision nodes, the most notable example, perhaps, being GERT (General Evaluation and Review Technique), a comprehensive network modelling package developed by Pritsker and Happ. These programs are not widely used in the construction industry. Their lack of use may be due to their cost, their complexity, or the reluctance of the industry to use anything other than the most simple deterministic networks. Discussions with practising planners suggest that a system which easily handled uncertainty

## THE USE OF UNCERTAINTY

would be welcomed as a way of quantifying risk but that the explicit inclusion of decision nodes is neither understood nor desired.

## The use of uncertainty

It is only by quantifying risk that decisions can be made in uncertain environments, and only by the recognition of uncertainty in planning can estimates of on-stream dates for important parts of multi-unit systems be made or direct knowledge of the risks of completion or non-completion be obtained. During the design process the planner must be aware of the accuracy of his data so that he can confidently choose the optimum action, because if the data are such that the range of solutions is greater than the difference between alternative solutions there is no basis for decision.

Possibly the major advantage that the consideration of uncertainty brings to planning is a sense of realism and a formalisation of the planners' 'gut' suspicion that some techniques lay claim to a totally fictitious precision. A case in point is the time/cost optimisation which is included in some planning packages and is taught on all construction management courses. This is a perfectly sound technique, of course, but like all design techniques should be used only by those who understand the limitations of the model to which it is being applied. A deterministic assessment using cost slopes of the optimum management action for the simple network illustrated in Figure 9.4 would result in the acceleration

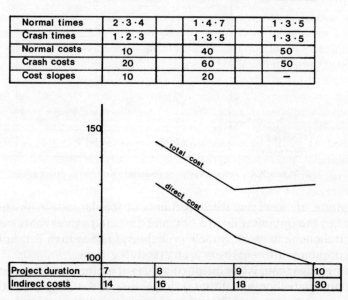

**Figure 9.4** The deterministic approach to time/cost optimisation.

# THE UNCERTAINTY OF CONSTRUCTION PLANNING

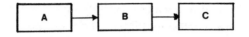

|  |  | A | B | C |  |
|---|---|---|---|---|---|
| Normal | mean | 3 | 4 | 3 | M ≥ 10 |
|  | variance | ·5 | 1·5 | 1 | M ≥ 3 |
| Crash | mean | 2 | 3 | 3 | M ≥ 8 |
|  | variance | ·5 | ·5 | 1 | M ≥ 2 |

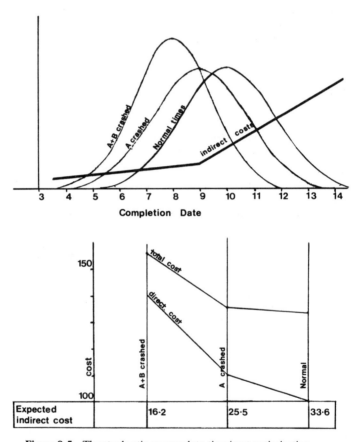

**Figure 9.5** The stochastic approach to time/cost optimisation.

of activity 1. If, however, the uncertainty of the duration estimates is carried forward to the optimisation process, and expected values (obtained by multiplying the marginal costs by their probability) rather than deterministic costs are used, then a different result is obtained (Fig. 9.5).

This simplistic case is not included as part of a proposal that expected values should always be computed, but as a warning of the danger of expecting too much from techniques which are based on artificially precise figures.

## The use of uncertainty in practice

This chapter has outlined the sources and types of uncertainty and the dangers of ignoring it. Techniques for evaluating it have been introduced. It is clear that advantage can be gained if methods can be found to incorporate the consideration of uncertainty into the networking systems which are used in the industry, but this advantage will only be gained if the techniques used are such that they do not make the programs difficult to use or the results incomprehensible.

# 10
# THE ANALYSIS OF UNCERTAINTY

The previous chapter has shown that in some circumstances the analysis of uncertainty as part of the general analysis process is desirable and that, at the least, the confidence of the planner in his results is increased if he has available the means to appraise risk. This chapter will review the two families of techniques which have been developed to handle the analysis of uncertainty and seek to provide pointers to their appropriate use in manual and computer-aided plan design systems.

### The two approaches

Statistics could be said to be the science of risk, and thus it is natural that a solution to the problem of handling risk should be sought there. The early network analysis packages did not, in general, provide facilities for the handling of uncertainty, but those which did adopted the PERT technique and used elementary statistics and a simple arithmetical process for the analysis.

PERT was written in the 1950s with one particular project in mind, the research and development required for the production of the Polaris missile. Those responsible for the Polaris project were realistic enough to realise that a deterministic approach to the planning of the project would not be satisfactory. The uncertainty involved in such a research-centred project is so great that it quickly swamps the analysis and makes deterministic forecasts literally incredible. To overcome this difficulty the developers of the PERT system built into their analysis not only a value for the expected duration of each activity but also a parameter representing the expected variation of the duration. These two variables were then carried through the analysis and all results were presented in terms of these parameters.

An alternative approach, which, because it relies on the power of the modern computer, was adopted rather later than PERT, is Monte Carlo Simulation. This technique uses the power (and patience) of the computer to analyse the project model many times, each time allowing the project variables to vary randomly within their allotted distribution. The results are stored as a list of recorded values on which statistical operations can be carried out. This

# THE STATISTICAL METHOD

technique has the advantage over the PERT technique of flexibility, but it buys this advantage at the price of computer aid.

## The statistical method

The techniques of statistics can be used to combine uncertain numbers, and these techniques were used by the developers of the PERT system to enable them to produce useful answers from data which, because of the nature of the research and development task, were very uncertain. Their method relied on the central limit theorem, which shows that the mean and the variance of the sum of a number of uncertain numbers can be easily predicted and that the distribution of the sum tends to be Gaussian in form.

Using the central limit theorem it is possible, if the value of the mean and the variance for each component is known, quickly to calculate the mean and the variance of the sum. Knowing the values of the mean and the variance, and knowing the form of the resulting distribution, the assessment of risk is trivial. The central limit theorem holds for any shape of component distribution and thus the approach used in PERT is applicable to the asymmetrical duration distributions for activities which are typically found in project planning.

PERT was built around the beta distribution as the standard for the component distributions. There is no theoretical justification for this choice, merely the pragmatic reason that it provides a wide variety of distribution forms and the calculations associated with it are very simple. The beta distribution is of the form

$$y = x^a(1-x)^b$$

which, depending on the values adopted by $a$ and $b$, produces a range of distributions both skew and symmetrical (Fig. 10.1).

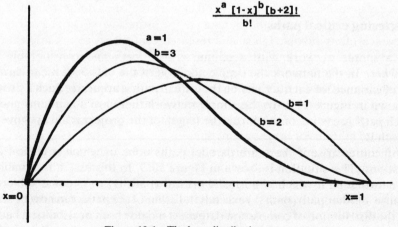

**Figure 10.1** The beta distribution.

## THE ANALYSIS OF UNCERTAINTY

The beta distribution has the advantages over other distributions of simplicity and flexibility. The distributions are defined in PERT not by the specification of statistical parameters, which would be foreign (and frightening) to most planners, but by the specification of three values of duration, the most optimistic estimate, the estimate of the most likely duration, and the most pessimistic estimate. The pessimistic and optimistic estimates are defined as the estimate of duration which, in the opinion of the planners, has a 10% chance of being optimistic (in the case of the pessimistic value) or pessimistic (in the case of the optimistic value). Denoting these values as $a$, $b$ and $c$, the parameters of the activity distributions can be calculated as

$$\bar{x} = \frac{a + 4b + c}{6}$$

$$\text{standard deviation} = \frac{c - a}{4}$$

Thus it is possible very quickly to move from estimates of duration, which are familiar ground to the planner, to the statistical parameters required for use in the statistical analysis. These parameters can then be used within the analysis of the network which, using the central limit theorem, can merely add the values of the activities on the critical path in order to produce the parameters for the distribution of completion date.

This approach is powerful and allows the planner even without the aid of a computer very quickly to produce distributions of event times and thus estimates of risk of overrun. The calculations required of the planner are simple and the planner is, for most of the calculation, working with numbers which are significant to him (an important advantage if errors are to be avoided). The approach does, however, have an important difficulty, the handling of parallel critical paths.

## Interfering critical paths

For a simple network with a unique critical path and considerable float elsewhere in the network the simple addition of the values of mean duration and of variance for each activity on the critical path is accurate. Such a situation is shown in Figure 10.2. In this simple network there are no circumstances in which path A can become critical; the length of the project is always governed by path B.

Difficulties arise if there are parallel paths none of which is critical in all situations. This situation is shown in Figure 10.3. In this case it is possible for the network path which is normally critical (path B) to become sub-critical because another path (path A) exceeds it. Either of the paths can now be critical and the distribution of completion date must include both possibilities, because durations of less than 36 days can be produced with either A or B critical and

# INTERFERING CRITICAL PATHS

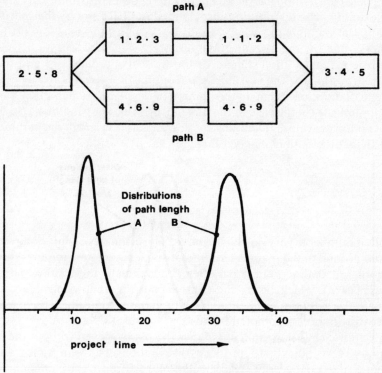

**Figure 10.2** A unique critical path.

each possibility must contribute to the probability of completion within a given time.

It is, of course, possible to handle this problem of interfering critical paths using the methods of classical statistics. For the case quoted the distribution can be shown to be given by

$$f(x) = f_A(x) \int (f_B((0 \to x)) \, dx) \qquad \text{A is critical}$$
$$+ f_B(x) \int (f_A((0 \to x)) \, dx) \qquad \text{B is critical}$$

but this calculation becomes onerous for large and complex networks where the number of possible interferences is much larger than the two described here. Figure 10.4 shows the result of the analysis of the effect of interference for the example in 10.3 and shows the way in which the presence of alternative critical paths increases the expected duration of the project.

The difficulties caused by complex networks to the PERT method make the search for alternatives to it necessary. This does not imply that PERT is not a useful tool, merely that in complex networks it loses the simplicity which is its major strength. One alternative to PERT, 'Monte Carlo simulation', has been

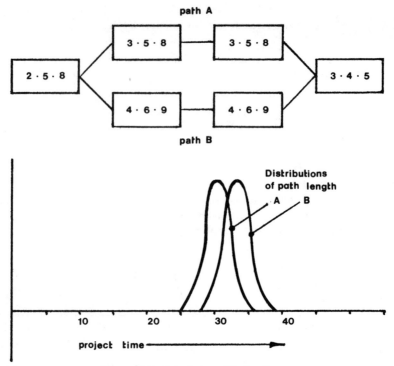

**Figure 10.3** Interfering critical paths.

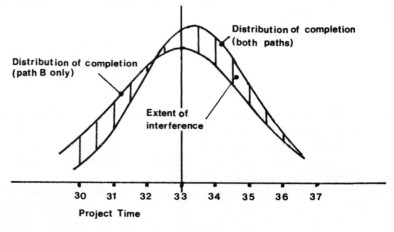

**Figure 10.4** The effect of interfering critical paths.

available for many years but is becoming increasingly attractive as the cost of computer power, on which it is dependent, is reduced.

## Simulation techniques

The analysis of uncertainty described above relies on carrying the parameters of uncertainty through the analysis and, using the methods of statistics, combining them to provide parameters of uncertainty of the results. PERT carries out this calculation using the same method as would have been used for a purely deterministic analysis, the analysis is done once only and the range of possible answers is represented in the result.

An alternative approach is to create the uncertainty of the result not by carrying uncertainty through a single analysis but by carrying out a large number of analyses using data values chosen at random from the original ranges. Thus the distribution comes not from the carrying of uncertainty through a single analysis but from the combination of the results of a large number of deterministic analyses. This alternative technique is known as Monte Carlo simulation.

It is obvious that the distribution produced by a simulation analysis is dependent upon the data values of the component parts. The analysis seeks to model many repetitions of the project and thus must take, as input, values of the variable (in this case duration) which are chosen from a menu of equally probable values. Thus in most cases the choice of the data values going into the analysis must consist not only of the choice at random of a value from within the specified range, but must also weight the choice towards the most probable values. This weighted choice is made by dividing the distribution of the variable into equal areas (that is areas of equal probability) and then by choosing at random between these areas. Figure 10.5 shows how, assuming a menu of ten values and a beta distribution, this would be done manually.

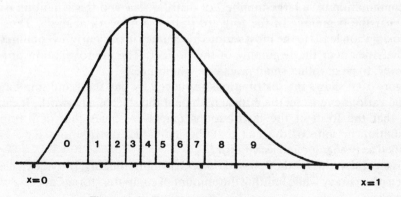

**Figure 10.5** Zones of equal probability.

## THE ANALYSIS OF UNCERTAINTY

When the distribution for each component data value has been specified and divided in this way a random choice is made between the equiprobable zones for each component and one cycle of the simulation analysis is carried out by allotting a value for the duration of each activity to that which represents the chosen zone. Many cycles are carried out and the results of the analyses combined to provide distributions of specified events.

### Data storage within simulation analyses

Simulation analyses will, of course, provide values for each event time in the network (early and late start and finish for each activity) for each simulation cycle. If a distribution is required for each of these events, or, more likely, if it is not known at the time of analysis which events are to be scrutinised, then all of these times must be separately stored for each simulation cycle. While this is possible, it is expensive either in computer power (if the data are stored in memory) or in analysis time (if the data are stored on disc). Such detailed storage is not necessary, however, except in special cases.

As has been noted previously the central limit theorem predicts that the combination of distributions will produce a distribution which tends towards the Gaussian. This property can be exploited in the storage and presentation of the data produced by Monte Carlo simulation. If it is assumed that the central limit theorem is true for each combination within a network, then each event distribution is Gaussian in form and can be reproduced by the specification of just two parameters, the mean and the variance. Consideration of the calculation methods for deriving these two parameters shows that they can be calculated from the cumulative values of time and time squared using the following formulae:

$$\bar{x} = \frac{\Sigma x}{n} \qquad \sigma^2 = \frac{\Sigma (x^2)}{n} - \sum \bar{x}^2$$

The assumption of normality is, of course, an approximation where the constituent distributions are not normal. The central limit theorem holds for the combination of 'a large number' of distributions and this condition is not met near the beginning of the forward pass of a network analysis. Thus the approximation leads to the most serious inaccuracy in the 'early' event times for the activities near the beginning of the project. The approximation proves, however, to be of rather small practical significance.

Figure 10.6 shows the distributions produced by one thousand simulations for the various events on the critical path near the start of a network. It can be seen that the form of the distributions quickly assumes that of a normal distribution in spite of the fact that the individual distributions have been specified as triangular and assymetric.

It is clear that this method, based on central limit thinking, provides a useful degree of accuracy while limiting the amount of computer storage necessary for the analysis.

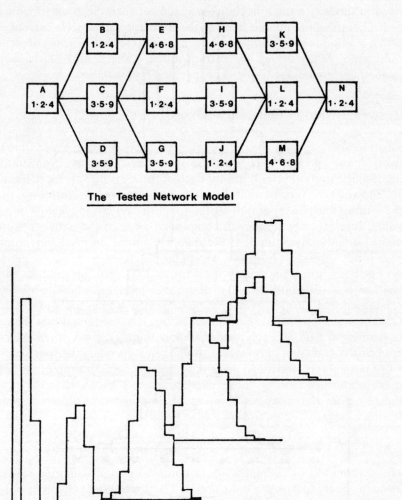

**Figure 10.6** The variation in distribution shape across the network.

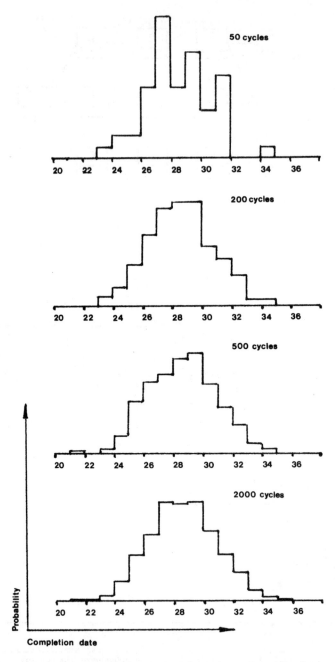

**Figure 10.7** The effect of the number of simulation cycles on distribution shape.

## The speed of simulation analyses

A second difficulty inherent in simulation techniques, particularly serious if they are to be used in an interactive environment, is that of computation speed. Simulation techniques as used in the past have usually included a very high number of cycles, perhaps 2000, 3000 or 5000 cycles being common. The work of Van Slyke (1963) showed that for high accuracy this was necessary. The project planner is faced with a situation where the highest levels of accuracy are not necessary but where, if the simulation analysis is part of a design system, itself a cyclical process, speed of response is very important.

If the analysis method is very time consuming, either it will not be used or the number of design cycles will be reduced because the operation becomes tedious, both of which are undesirable. Figure 10.7 compares the distributions for the completion date of the network shown in Figure 10.6 with distributions produced using 50, 200, 500 and 2000 simulations. Clearly the distributions produced by 50 and, to a lesser extent, 200 simulations are unsatisfactory as they stand and could not confidently be used as the basis for further calculations in such techniques as risk analysis.

If, however, the normal distributions predicted by the central limit thinking described above are assumed, then the differences between the answers become much smaller and it is clear that for most purposes satisfactory results could be obtained using only 50 or, at the most, 200 simulation cycles. It seems that the use of very large numbers of simulation cycles is wasteful and may be harmful in that it produces an 'accuracy' which is purely illusory.

The result obtained from the comparison described above is to be expected, for the standard error of the mean and the variance of a sample diminishes as the square root of the sample size, thus following a curve which shows rapid gains in accuracy for small sample size, but which requires very large increases in sample size if comparable gains are to be made for large samples. This is illustrated by the curve in Figure 10.8.

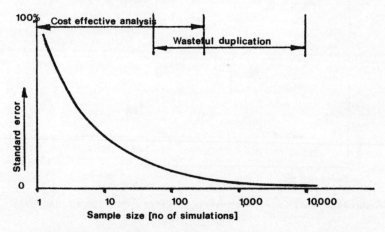

**Figure 10.8** The diminishing returns of multiple simulation cycles.

It seems that for normal use the speed of the simulation process can be dramatically increased (by a factor of between 40 and 100) by accepting the inaccuracy of the assumption of normality. This inaccuracy is of small practical significance and its acceptance makes the use of simulation techniques within interactive design an attractive proposition.

**Simulation – a viable design tool**

Monte Carlo simulation is a powerful tool for the handling of uncertainty and is particularly useful in those situations where the presence of interacting critical paths makes the use of PERT inaccurate or cumbersome. Simulation does, however, require thoughtfully produced computer programs, and for handling uncertainty, PERT remains the only method available to the planner who does not have computer support.

The difficulties inherent in the use of Monte Carlo simulation – that is, the need for computer power and the slow analysis speed – can be much reduced by using the insights provided by the central limit theorem, and the inaccuracy introduced into the analysis by adopting techniques based on these insights seems to be of little practical significance.

# 11
# THE SIMULATION OF UNCERTAINTY

**The viability of simulation**

It has been shown in the previous two chapters that the consideration of uncertainty is desirable if the planner is to have confidence in his design and if he is to make realistic statements about the extent and the cost of risk. Techniques have also been suggested which, at the cost of an introduction of a small amount of inaccuracy, can increase the speed of the simulation process and reduce the size of the necessary hardware. This chapter will show that the hardware necessary for this type of analysis is now widely available and will describe a program which will carry out the analysis using graphical input and output of data.

One of the problems with the simulation analyses which have been available in the past to project planners is that they have required considerable computer power. This power is necessary if the simulation is to be done in a short time, for if time is not a constraint even a manual simulation analysis could be done. In practice time is always an important constraint, and it is of crucial importance if the analysis forms part of an interactive system, for interaction becomes tedious if the response time of the computer is very long. Computer power reduces the time required for analysis by providing faster calculations (typically a powerful computer will be operating in a compiled language) and also by minimising the transfer of data into and out of memory from and to disc.

Until recently the computers with the power necessary to carry out simulation analyses on large networks in a reasonably short time were available only in relatively inaccessible form, often as centralised mainframe machines set up for, perhaps, the handling of accounts and payroll. Such machines are not usually convenient for interactive work either in architecture, location, or organisation, but recent advances in computer design have made their use for this purpose unnecessary.

The latest generation of machines suitable for office installations have the power and capacity necessary for simulation analyses on large networks (that is between 200K and 500K bytes) and have very good high-resolution graphics facilities, permitting the use of the graphical I/O techniques which have been described previously in this book. They are often single-user machines designed to be operated and housed at the place where the user usually works. Thus they are the ideal tool for interaction as they can be used as part of the designer's normal work environment alongside the other aids, books, files, drawings, etc. which he would normally have available. This type of machine is ideally suited to simulation techniques, and its availability has changed these techniques from being an academic plaything to being a viable and useful tool.

## The difficulties of simulation

As has been noted above, a simulation analysis is a useful part of an interactive design system only if it is fast. This need for speed must be recognised in the writing of the computer programs and at times it may be necessary to sacrifice accuracy on this particular altar. The previous chapter argued that savings in calculation time could be made without serious loss of accuracy by reducing the number of simulation cycles, and by storing cumulative rather than individual values of data. A third area of saving is the generation of the individual distributions prior to the selection of deterministic duration values. The implications of these recommendations to the development of a computer program must now be considered.

### *The necessary number of simulations*

There is a sense in which, in using this type of analysis, you get what you pay for. If the analyst wishes accurately to predict the distribution of any event in the network then he will require a very large number ($>2000$) of simulation cycles to produce it. Reducing the number will, as was shown in Figure 10.7, produce a progressively less satisfactory result. If, however, the analyst requires an estimate of mean event time and is satisfied with an assumption of a normal distribution and an estimated value of standard deviation, then the number of simulation cycles can be enormously reduced. The previous chapter showed that in these circumstances 50 or even 20 simulations would produce answers which were sufficiently accurate for most planning purposes. Obviously the analysis time is directly proportional to the number of cycles used, and the program developer who insists on a large number of cycles must constantly remind himself of the cost in analysis time of this policy and of the limited utility of the accuracy he is producing for his user.

### *Data storage*

Even in a large computer the problem of data storage can be troublesome if both the number of events in a network and also the number of simulation cycles is high. This problem can be overcome as shown in the previous chapter by combining the data as they are produced and carrying forward a cumulative total. This technique, bringing the advantage of reducing the amount of

## THE DIFFICULTIES OF SIMULATION

memory required or of increasing the speed of operation of the program, has the disadvantage that, like the method described above, it destroys the shape of the distribution and imposes a normal distribution on the results. The previous chapter showed that this is not a serious error (see Fig. 10.5) at least for events away from the early events near the start of the network. As the assumption behind this technique is exactly the same as that behind that described above and as it brings the same inaccuracy, it would be foolish to use one without the other. Storing the data in the way suggested here has a secondary advantage in that the data are in a form which makes them very easy to present or to use in following programs (time/cost optimisation or risk analysis).

## Distribution calculation

It has become usual to represent the distribution of activity duration by a beta distribution given by the formula

$$y = x^a(1 - x)^b$$

There is no theoretical basis for this choice; it is chosen merely because the curve can be moulded, by the selection of the parameters $a$ and $b$ to fit the curve, to a shape that experienced planners would draw to describe the distribution of durations. The curve can be symmetrical or asymmetrical, flat or sharp. Where manual calculations are being carried out, particularly within the PERT program, the distribution provides easy formulae for both mean and standard deviation and is thus convenient.

When the method of analysis is changed to a Monte Carlo simulation such as is being described here, and when the means of doing the work is moved from the manual environment to the computer, then the beta distribution becomes rather less convenient. No direct use is now made of the mean and standard deviation of the component distributions; instead the centroids of the zones of equal probability must be identified.

Clearly this calculation does not present any difficulty: given the three duration values as is usual in PERT (and assuming a value of $b$) the value of $a$ can be calculated. Progressive numerical integration of the curve so specified can then be used to produce the positions of the various centroids and these can be used as the menu items from which the random choice is made prior to each simulation cycle. While this calculation is simple, it is also cumbersome and time consuming, especially when it must be repeated for each activity for each simulation cycle. There is merit, therefore, in identifying a distribution which allows the direct calculation of the centroids of the equiprobable zones.

It has been noted in the previous chapter that the results of simulation analyses are insensitive to the shape of the individual activity duration distributions if the number to be combined is greater than two or three, and that this condition applies to all the events in the network apart from those very near the beginning of the forward pass. In view of this insensitivity it seems to be

# THE SIMULATION OF UNCERTAINTY

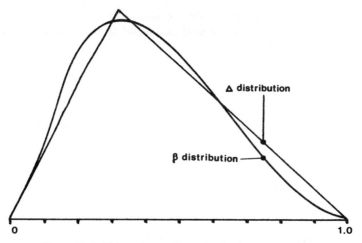

**Figure 11.1** Triangular and beta distributions compared.

unnecessarily pedantic to insist on the use of a traditional but not theoretically derived beta distribution, for any distribution which approximates to the general shape of the expected distribution could be used. When this is considered along with the need for a geometrically simple shape, it seems that a triangular distribution could form a compromise between simplicity, giving fast calculation times, and shape, giving a realistic representation. Figure 11.1 contrasts the triangular distribution recommended here with the beta distribution it replaces.

These three techniques, each of them simple, provide the means of greatly increasing the speed and thus the usefulness of Monte Carlo simulation. They have been used along with the graphical I/O techniques developed earlier to produce a fast and pleasant Monte Carlo-based planning program.

## A simulation program

If simulation techniques are successfully to form part of the planning process they must be fast and convenient to use. By combining the three elements described above, that is by reducing the number of simulations to the minimum, by storing the results in the most efficient way possible, and by simplifying the calculations to be performed within the analysis through the use of triangular distributions, an analysis program can be produced which is fast enough to form part of an interactive design system. By using the graphical techniques which were developed for the program described in Chapter 7 of this book such a program can be easy and pleasant to use.

Although the program to be described is written for one machine, the principles behind the program are, of course, universally applicable to machines which support large memories. It is probably inevitable that programs

# A SIMULATION PROGRAM

which concentrate on I/O, as do the programs described in this work, should be relatively machine specific, for much of the software is written around the hardware which happens to be available.

## The analysis routine

The analysis algorithm used in the simulation program is the same one that is used in deterministic programs. The durations of the activities are selected at the start of each simulation by a random choice between equiprobable values, and the chosen values are rounded to the nearest time unit. The selection process includes the recalculation of the equiprobable zone centroid in each simulation cycle, because the calculation using triangular distributions is so fast that there is no advantage in setting up an initial list of possible durations and storing it, as would have been done if the more complicated calculation procedure implied by the beta distribution had been followed. The duration values are used in the analysis in the normal way. The following algorithm shows how the durations can be calculated:

```
var t,u,v:real
begin
    u:=(c−a)/(b−a);
    v:=random+0.5
    if v<u then t:=a+sqrt(v*(b−a)*(c−a));
    if v>u then t:=b−sqrt((1−v)*(b−a)*(b−c));
    d:=round(t);
end;
```

A default value of 50 for the number of simulations has been set in the program, and although this can be altered by editing the program it is felt that this should not normally be necessary and is not desirable. The cumulative values of time and time squared for each event time are stored and carried forward to the stage of the presentation of data.

The program as presented, running on an ICL PERQ, performs analyses at a speed of approximately 15 ms per activity: that is, less than one second per activity for the whole simulation process if 50 cycles are used. Thus a relatively large network (of 400 activities) will require around five minutes for its analysis and the results of that analysis are stored so as to be continuously available thereafter for reference.

## Data input

The format chosen for data input is the same as that which has been successfully used earlier. The only special data input routine required is the duration histogram which now must present not a single-value duration, but a three-value distribution. In order to do this the display is altered as shown in Figure 11.2 so that rather than producing bars the length of which represents the

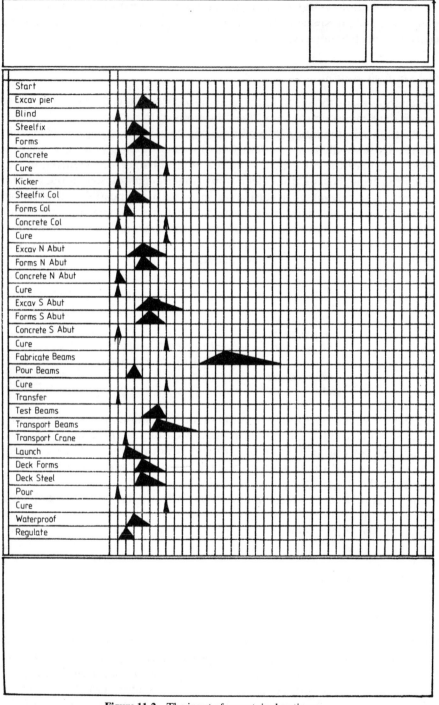

**Figure 11.2** The input of uncertain durations.

## A SIMULATION PROGRAM

deterministic duration, a representation of the triangular distribution is plotted on the histogram grid.

For this program use is made of the four-button cursor available on the digitiser being used, three buttons being used to indicate the optimistic, likely, and pessimistic durations respectively, but in other machine configurations it may be helpful to receive three values and have the program sort them into the required order. Scrolling in the horizontal and the vertical directions is available.

This graphical display, like its deterministic equivalent, has proved to be both pleasant to use and highly efficient in that gross error is virtually eliminated. The triangular distribution used as part of the display is almost completely self-explanatory and thus the need for documentation is minimised.

## *Data output*

Experience has shown that Gantt charts are a most efficient means of representing planning data for deterministic analyses. The output of simulation programs, however, contains one extra dimension, that of uncertainty, which is not easy to represent on a bar chart. Various options for presenting this dimension are available, and many have tried means of showing distributions used in the duration histogram, that is the drawing of a distribution, and applied it to the Gantt chart. This was found to be most unsatisfactory because each line of the Gantt chart supports four events, the early and late start and finish of the activity, and drawing the four distributions in the limited space available produced confusion and not much else (see Fig. 11.3)! A second option was to represent the activities by bars, as in Chap. 4, but to shade the bars according to the probability of the activity taking place at the indicated time. This method produced a more easily read chart than the first option had but suffered from the disadvantage that it showed only the left-hand side of the start and the right-hand side of the finish (see Fig. 11.4). It was felt that neither of these options was satisfactory, not least because they complicated the Gantt chart, the strength of which is surely its simplicity.

This problem of the representation of uncertainty on the Gantt chart can be solved by providing optional information about the probability of any event in

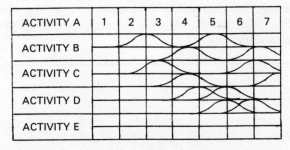

**Figure 11.3** Uncertainty plotted as individual distributions.

# THE SIMULATION OF UNCERTAINTY

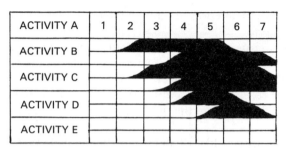

**Figure 11.4**  Uncertainty represented by shading.

the network through the use of the cursor and the windowing facility of the display. The Gantt chart is drawn using the mean event times as calculated in the simulation analysis. If the operator wishes to have information concerning the distribution of event time for any point in the project he uses the cursor to indicate the point and a distribution of event time against a horizontal axis annotated in calendar dates is plotted. This facility is shown in Figure 11.5.

This solution to the problem has proved to be effective and pleasant to use. In particular it preserves the chief virtue of the Gantt chart for data presentation, namely its simplicity. It also brings the considerable advantage of offering a program which has Monte Carlo simulation at the heart of its analysis routine but which appears to be simply deterministic. The display does not show uncertainty unless interrogated and thus a program of this type can be used at various levels of sophistication.

## The simulation program in operation

The program is very fast, it performs the analysis of a 250-activity network in under three minutes, and produces results of an accuracy sufficient for most planning purposes. It achieves this by including in its analysis the 'central limit thinking' developed in the previous chapter. The output of the program is in the form of mean times and standard deviations for each event in the network or of screen-displayed Gantt charts. The numerical results are in a form which is readily usable by following programs. The speed of the program is such that it seems that there is little advantage to be gained by the use of a deterministic equivalent other than familiarity with the concept, comfortable though inaccurate, of fixed durations.

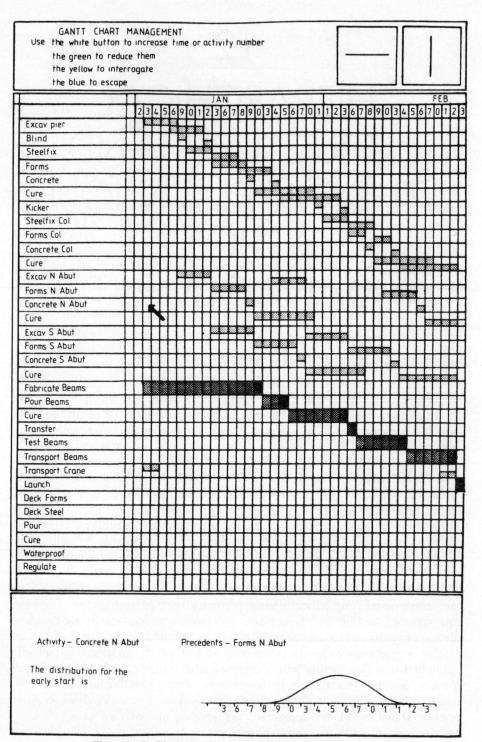

**Figure 11.5** The interrogation of a stochastic Gantt chart.

# 12
# HIERARCHICAL STRUCTURES AND THEIR APPLICATION TO PROJECT MANAGEMENT

**Projects and complexity**

A project is a complex task. It is this complexity of the task which is distinctive in the project and not either the objective of the task or the environment in which it is performed. The project can be concerned with pure research, with construction, with design, or with the organising of a church social; it is the element of complexity which is the common factor.

If complexity is to be managed it must be anticipated and the possible decisions concerning its management and their outcome should at least be considered prior to the events which lead to them. It is for this reason that forecasting and planning are of paramount importance in project management; decision making is a lottery unless techniques are available to handle the complexity inherent in the project and to enable the planner to anticipate his difficulties.

With increasing complexity, which makes planning and forecasting important, comes increasing difficulty in the planning function itself and thus we have the ironical situation that those projects in most need of planning are the most difficult to plan and so the ones which are often planned least effectively.

The computer provides the planner with a tool which can handle complexity without pain. This facility was recognised early in the development of computers, and for this reason their use was adopted enthusiastically in the construction industry and elsewhere. Too often, however, the computer merely transformed the complexity without significantly reducing it: complexity was fed into the computer and complexity came out of it. The decision making process remained difficult, particularly as the output of the computer was often in acutely indigestible form, and the use of computers as aids to project management decision making has never reached its potential level.

Reference was made earlier in this book to Asimow's definition of design. Project planners may, however, prefer Jones' definition – 'a very complicated act of faith.' Planning is a design process and, as has been shown, insight into it can be gained by examining experience of design in other, more conventional areas of design.

In addition to being iterative, design is hierarchical, proceeding from broad concepts to fine detail. The adding of the detail to a design continues until the required definition of solution is achieved. The level of definition which is appropriate will depend on the reason for the design activity, product design carried out for a feasibility study need not be as precise as that required for manufacture, and in the field of project planning the definition required for cost estimation is significantly less than that required for the day-to-day running of the project. Because the design process is both iterative and hierarchical a three-dimensional diagram is needed to represent it; such a diagram was shown in Figure 2.3. As each level is satisfactorily completed the design proceeds to the next level down, and this progression continues until either the required definition is achieved or, following Lichtenburg (see Ch. 9), further detail is too uncertain to be useful.

The need for a progressive increase in complexity, that is for progress from the simple to the complex through the addition of detail, arises from the need for decision making within the iterative design loop. The decision maker must reduce complexity if he is to make decisions concerning it. Details of second order must be excluded from the first order decision making process and must be included only when the first order decisions have been made. Obviously this simplified picture of an orderly progression through a hierarchy can be dangerous, and a good designer constantly reviews his assessment of which factors are of primary and secondary importance and the effect of the lower-level decisions on the high-level decisions already made. It must be said that this reassessment is painful and costly, for the designer develops an affection and a loyalty for his design which makes rational self-criticism difficult.

## Plans as hierarchies

The design process, as has been noted, makes use of a hierarchical structure. Detail is progressively added, increasing the definition of the final design and reducing the uncertainty of the parameters (cost, weight, time, etc.) which are associated with it and which may constitute criteria of success. For most products this hierarchy is a means to an end and will be visible in the final design only as a convenient basis for the grouping of information, such as is the case in the content of the working drawings. For project planning, however, hierarchies are a fundamental part of the structure of the product. Plans are hierarchical in nature and will be used as such throughout the project.

## HIERARCHICAL STRUCTURES

A plan is a means to an end, not an end in itself. It may be used for the control of the project (and there is good cause, as shown in the first chapter of this book, for regarding control as a continuation of the planning phase within project time), or it may be used for management purposes prior to the start of activity, for forecasting or organisation. In either case the degree of detail required of the plan depends upon the use to which it is being put. Thus the managing director of the company, requiring a progress report for a construction project, wishes to know only in the broadest detail how much work has been done and how much should have been done. The site foreman, on the other hand, needs to know the fine detail of what is happening in a very limited part of the project. Both of these individuals are using the same plan but their method and their requirements are very different. For project plans, therefore, the hierarchical structure is important not only in design, as in all design disciplines, but also in the use of the plan.

### The use of hierarchies to handle complexity

In order to handle complexity the project planner/manager must be able to group information for decision making purposes. Of the various types of grouping which could be used, a hierarchical structure is of the most interest because it appears, apparently independently, in several aspects of the manager's work. The concept of hierarchical structure is implicit in the design process used to formulate the project plan, the conventional way of representing the organisation of the project is hierarchical, and the information flow and information requirements within the project are similarly hierarchical in nature.

### The hierarchical nature of design

As has frequently been remarked, design is a cyclical process. The design process represented by Figure 2.1 consists of repetitions of the sequence idea–test–modify, at each stage testing the design against some measure of acceptability. Solutions which are less acceptable than previous ones are discarded, solutions which are more acceptable are retained and, if possible, improved.

For the simple decision it may be possible for the decision maker to remain at this top level of detail. Whether or not to have a cooked supper, or which television programme to watch, can be decided at this top level without consideration of further detail. However, most decisions with even a small degree of complexity involve a hierarchy of detail. A decision to buy a car is followed by decisions concerning the number of seats, the colour, etc. Thus the decision takes the form of an inverted tree (Fig. 12.1).

Without this tree structure the decisions would be much more difficult to

## HIERARCHICAL NATURE OF DESIGN

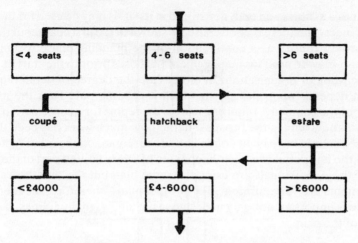

**Figure 12.1** An hierarchical decision structure.

make, for a multilevel hierarchy such as the one shown contains very many possible decisions. By considering the decision in this ordered way the decision maker is faced, not with complexity, but with a string of relatively simple decisions, and a rational choice can be made more easily than would otherwise have been possible.

Although this decision procedure appears to be very orderly, it should be noted that the formal descending path is frequently not followed. It may be, for example, that in the example of car selection the cost of the available models is not as hoped, that the decision on the number of seats has to be changed, or the desired colour is not available. Thus decisions have to be reversed from the lowest level and at the last minute. There is always the possibility that a lower-level decision may invalidate an upper-level decision and that the decision process must be repeated.

The design process is similar in structure to the decision process described above, the difference being that each level of the structure is now the cyclical process of design, and the designer moves to a lower level only when the design process is satisfactorily completed at the upper level. Thus the design process, while retaining the tree structure, retains it in three dimensions as shown in Figure 2.3.

This hierarchical structure of design is familiar to designers working in all fields. The bridge designer proceeds to decide on beam sections only after deciding on the material for them; the roadworks engineer considers drainage after choosing alignment. But, as these examples show, the need to reverse the process, for the drainage to influence the alignment or for the shape to affect the choice of beam material is very strong in the design process. Very frequently the designer defers the upper-level decision until lower levels have been considered. He considers alternative designs and alternatives within the alternatives, and pursues these paths until they give him sufficient data on

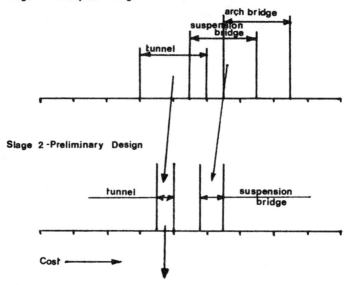

**Figure 12.2**  The increasing accuracy of progressing design.

which to base the upper-level decisions. Thus the descent through the design hierarchy fulfils two purposes: the provision of data for upper level decisions and the provision of detail for the specification of the design product.

Because the purposes of descent through the hierarchy differ, so do the levels to which the descent must go. For the making of upper-level decisions the designer needs only to descend to that level which gives him sufficient data confidently to make these 'broad brush' decisions. The designer adds detail in the first stages purely to help him eliminate some of the high-level possibilities; the adding of detail prematurely would waste his time. Figure 12.2 illustrates this point for a river crossing. The addition of detail tightens the range of the forecasts (in this case of cost), removes the overlap which occurs between the original cost estimates at the stage of conceptual design and takes away the difficulty of deciding between the three alternative courses of action for development at the preliminary design stage.

For the provision of detail for specification purposes the designer probably needs to go to greater depth in the hierarchy, but the level to which he can usefully go depends upon the uncertainty which he can handle and process. Structural steel can be designed and detailed to extreme detail, but project plans, although designed by identical processes, cannot be pursued to the same degree of detail during the planning stage; such detail must be delayed until further detail becomes available during the control phase.

Lichtenburg has shown that, in the related field of cost estimation, the addition of detail beyond a certain point is futile. In the same way uncertainty provides a cut-off point beyond which the adding of detail to a project plan may

## THE HIERARCHY OF ORGANISATION

be not only a waste of time but actually harmful, for unwarranted certainty in planning detail can make the plan unnecessarily rigid at an early stage, causing it quickly to become obsolete, and reducing the respect the users have for it. This second use of the hierarchy is limited, therefore, by both the need for detail and also the availability of data. In project planning the data become available only as the project proceeds, and thus the level of detail which is available increases as the activity approaches and the need for detail increases, leading to the distinctive information wedge which has been described previously.

It is clear from this discussion that the design of complexity is hierarchical. If a computer-aided design system is not similarly structured the designer will produce the hierarchy informally, either using the computer independently at each decision stage or, as is frequently the case in project planning at present, introducing the computer into the design stage only at the end of the process when all the decisions have been made. This latter use is analysis, not design. Any computer-aided *design* system must take the hierarchical structure of plans seriously and its developer must realise the ubiquity of these structures in the project environment and understand the way in which the various hierarchies interact. Any planning system which is not hierarchical is, therefore, modelling only one part of the design process. The hierarchy is there whether the programmer includes it in his program or not; if he does not include it then there is a limit to the amount of the design process carried out by the man and machine working closely together.

## The hierarchy of organisation

Management is necessary whenever a task is larger than can be accomplished by one person, and it thus implies organisation, the splitting of the task. This division of the task implies a hierarchy of responsibility if control of the task is ultimately to rest with one individual. Superficially this would appear to lead to a two-level hierarchy consisting of the manager at the top level and everyone else at the second level, but in practice such an organisation is not feasible when the management task is large. The management function becomes too big for one individual and must itself be split, restricting the number for whom each manager is directly responsible to between five and twelve, depending on the nature of the work. A multi-level hierarchical organisation is thus formed in which the number of people supervised by one individual (the management span) is limited. Figure 12.3 illustrates how a large number of individuals can be supervised without the use of large spans through the use of hierarchies.

Responsibility for large tasks, of which projects are an obvious example, is thus inevitably hierarchical. The need for information and for decision making exhibit the same structure as has been seen previously; detailed decisions feed on detailed data available at the lower levels of the hierarchy, and such decisions are made at these levels.

# HIERARCHICAL STRUCTURES

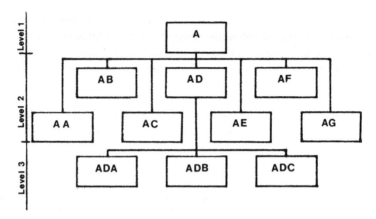

Figure 12.3   The reduction of management span through the use of hierarchical structure.

When this pattern of decision making within the organisation is superimposed on the pattern of decision making in time which was discussed above (that is that detailed decisions are deferred until near the date of the action, when detailed data are available), it will be seen that the majority of project decisions (although admittedly of tactical rather than strategic scope) are made at a late stage by those inhabiting the lower levels of the managerial hierarchy.

These decisions are not the major decisions of policy, which have probably been made at the planning stage, but the minor decisions concerning the positioning of plant, the deployment of labour, or the ordering of materials. Individually these decisions are of little importance, but collectively they dictate the success or failure of the project and it is important that those who make these decisions should have the same access to the information they need as their more senior colleagues.

Although the structure represented by Figure 12.3 is found in all large organisations, the diagram is a dangerous oversimplification of the actual situation. The source of the danger lies in its ignoring the possibility of communication between people at the same managerial level; it assumes that all information flow is vertical, whereas in practice there is a constant horizontal flow in most organisations. It is dangerous in that in formalising the vertical it formalises an individualisation of decision making. Likert has shown that rather than formalising the individual and his relationships, it is possible and helpful to formalise the group, emphasising the part the group must play in decision making within the organisation.

The manager is now the 'linking pin' between groups, he is the leader of the subordinate group and a voting member of the superior group. He acts, not as an autocrat who feeds on the data provided by blinkered subjects, but as a coordinator and arbitrator.

Experience suggests that the construction industry has, at least in its better-managed projects, widely used this management structure without knowing that it was doing so – its use has been informal. Daily section meetings over coffee and weekly site meetings in the more formal setting of the agent's office are the norm; the manager–man linkage of the formal organisation chart is not usually adhered to, for the good and sufficient reason that, as the behaviouralists would expect, people respond well to the taking of responsibility.

The traditional pattern of organisation has, however, many attractions for the creator of CAD/CAM systems. Its simplicity, the uniqueness of its linkage, the unidirectional information flow, all have much to commend themselves to the developer of computer systems. But the developer must be careful that in assuming the traditional manager–man linkage in his system he does not preclude the type of group decision making to which Likert refers and which is at present, in the absence of computer 'improvements', serving the industry well.

**The flow of information within a hierarchy**

The decision maker wants information in carefully controlled quantity and carefully specified quality. The presence of too many data in too fine detail swamps him in complexity; the lack of sufficient data oversimplifies the decision and leads to suboptimal performance. It is therefore evident that, as in other aspects of project management, there is a hierarchical structure present in the management of data. Thus the senior management of the company are interested in overall performance, the site manager in the progress of sections of the work, and the section foreman in the performance of individual gangs or workmen.

Obviously, following the principles of management by exception, senior management may wish to see the detail of some particularly critical operation, or of one which is persistently falling short of some pre-set objective, but they can only identify their need for detail if they can grasp the overall picture. To enable them to do this the level of detail must be suppressed.

Unfortunately, as in the case of the organisational hierarchy, this picture of a tree structure within a project is a simplification. Decision making may require other information than is provided by a formal tree and there will be need for horizontal as well as vertical flow. The structure showing the need for information may, therefore, be much more like Likert's model of organisation than the simple tree of the traditional structure; there is a much more diverse flow.

Again the responsibility of the creator of the planning and control system becomes apparent. If he is not aware of the need for horizontal flow of information, or if he chooses to ignore it because it complicates what otherwise would be a neat tree structure, he risks imposing an inadequate access to information upon the project team. In these circumstances the team will either be less effective than it was in the non-computer environment when these artificial paths did not exist, or it will gather information informally, as it has in the past, and bypass the computer system, rendering the computer system redundant and removing the advantages which the computer potentially could bring.

## The application of hierarchical concepts within computer systems

This discussion has sought to show that hierarchies are an inherent part of the handling and discussion of complexity. It has further sought to outline the dangers of ignoring either the complexity or its structure in the design of computer systems and the possibility of combining the various hierarchical structures in the project to form the basis of a project management data system. If the hierarchical nature of decision making, of organisation, or of information needs are ignored when the model of the project or the I/O system are being developed, then the model will be used only as a tool of secondary importance and there is a danger that extensive work on CAD systems for project design will be found to have been wasted.

# 13
# THE HIERARCHICAL DIVISION OF PROJECTS

## Activity grouping

The previous chapter has shown that projects are hierarchical systems and that it may be advantageous to reflect this structure in the model of the project used for planning and control. If this is to be done then some thought must be given to the way in which the division of the project at each level is to be carried out, because hierarchical structures, by their nature, have intermediate stages between the whole and the atom, and there exist alternative methods of defining the elements at the intermediate stages.

A unilevel plan does not encounter this problem of intermediate division. The program is divided into working units – the tasks which comprise the project – which, being indivisible from the planners' point of view, select themselves. A multilevel plan, on the other hand, while offering no choice at the top level (the whole project is one unit) and the bottom level (the individual tasks), requires that intermediate level division be defined by the planner. These intermediate groupings can be based on either functional or divisional organisation concepts and their choice can significantly affect the model of the project and its use.

## Divisional grouping

Divisional grouping is the splitting of a complex task into a series of simple tasks grouped according to location. This organisational structure is frequently used in the construction industry, and it can also be used as a basis for the grouping of intermediate levels in a hierarchy. Thus, for a motorway construction project, the task may be split into sections comprising lengths of road; for a house-building project streets may form the subdivisions.

Although the sections of a civil engineering site are often chosen in terms of geographical location and apparently with some degree of arbitrariness, in fact the boundaries between sections are often carefully considered and fixed such

## HIERARCHICAL DIVISION OF PROJECTS

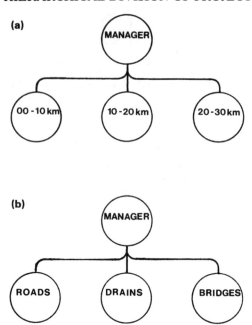

**Figure 13.1** The alternative organisational models; (a) divisional grouping; and (b) functional grouping.

that the section is as autonomous as possible. Thus the section of the motorway would include its own drainage outfalls, the whole length of sideroads, etc. and would exclude elements of work which were logically tied solely to work on other sections, such as the diversion work to a river upstream of a river crossing. Figure 13.1a illustrates this type of organisation.

The implications of this method of choosing the boundaries between elements in hierarchies are interesting, for the search for autonomy can be restated in terms familiar to network planners as a statement that as far as possible the tasks within a section should be chosen to minimise the number of logic links passing between sections.

It is clear that this ideal of autonomy is never completely met. Frequently a task downstream is delayed by problems to upstream tasks in another section of the job; a task cannot be located in two sections at the same time. But the fact that the ideal of autonomy is the basis of the division of projects gives a pointer to the implications of the use of divisional grouping for the intermediate levels of hierarchies; it seems that this type of grouping will seek to minimise cross linkage due to logic.

### Functional grouping

The alternative method of organisation for a project is known as functional grouping. Functional grouping consists of putting within one section the tasks

## MATRIX GROUPING

which are similar in skill or the type of productive resource required. The motorway scheme may thus have sections which consist of the whole of the earthworks, the structures, or the blacktop; the housing scheme may have a bricklaying or a joinery section. Figure 13.1b shows this in diagrammatic form.

As in the case of the divisional organisation, the functional organisation is designed so as to make each section as autonomous as possible. In the case of the divisional structure this was achieved by minimising the number of logical dependency links which crossed the section boundary. In the case of a functional organisation the number of boundary crossing links is again minimised, but in this case it is the resource-based links which are given the primary attention of the designer. Thus the earthworks section might include within its area of responsibility the maintenance of haul-roads, which are not strictly within its terms of reference but which rely on the resources of the earthworks section for their completion.

Again there are implications for the hierarchically organised model, for a grouping of activities in accordance with this functional pattern minimises the resource links which cross the boundaries between the limbs of the hierarchical tree.

## Matrix grouping

A combination of divisional and functional grouping is possible and is favoured by some managers. In this organisational model some degree of control is exercised both divisionally and functionally by making each section responsible to both sectional and functional managers.

The motorway construction project which was described by Figure 13.1 can, therefore, be grouped as shown by Figure 13.2, the divisional manager

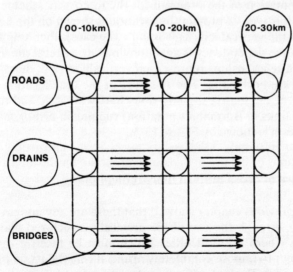

**Figure 13.2**  Resource flow in a matrix organisation.

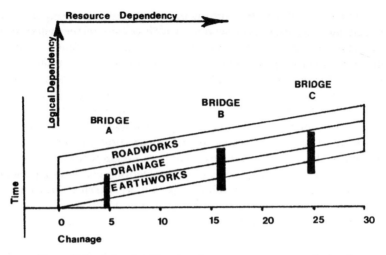

**Figure 13.3** A time/location chart for a motorway construction project.

coordinating the activities within his 10 km section of the work, but liaising with his functional colleague for the organisation of the particular specialist tasks. It is interesting to note that the separation of logic and resource links which became evident during the discussion of divisional and functional organisation can now be seen in the diagram of the organisation matrix. Figure 13.2 shows the lines of responsibility within the project organisation and it is clear that most questions concerning resources move along horizontal lines, whereas logical links are, for the most part, vertical.

If, as is often the case, a variant on the familiar 'line of balance' chart is used for the representation of the program for the motorway scheme (the variant being that for this type of project the location is shown on the horizontal axis rather than on the vertical axis as is usually the case), then this separation of logic and resource dependency is again obvious. In general the logic links are concerned with relationships represented vertically on the chart, whereas the resource links are horizontal. The structure of the time/location chart is thus similar to that of the matrix organisation chart with logic and resource links (hard and soft links in Burmann's notation) running in orthogonal directions. This phenomenon is shown by Figure 13.3.

## The occurrence of cross links in hierarchical plans

Although the previous chapter showed that there are advantages to be gained by the planner through the use of hierarchically structured plans during his design process, there are difficulties which must be faced if the concept of hierarchies is not to impose completely artificial constraints on the planner as he builds his model. The first difficulty, which has already been discussed, is the

## THE OCCURRENCE OF CROSS LINKS

grouping of the elements at the intermediate levels of the hierarchy; the second is the occurrence of cross links which can, by robbing the structure of its simplicity, cause it to lose some of its appeal.

The major reason for the introduction of hierarchies into planning systems is the ease with which they handle complexity. The previous chapter showed how, using hierarchies, the designer can concentrate on adding detail to those areas of his model which require it, confident that he will not lose sight of the behaviour of the model on the large scale, and the controller can access the model at whatever degree of detail is appropriate to the task in hand.

The creation of simplicity out of complexity depends upon the ability to separate sections of the work. If there are no cross links between the lower levels, then this separation is complete (Fig. 13.4a). If cross links occur then separation is not possible without serious error (Fig. 13.4b).

It is possible for simple hierarchies such as shown by Figure 13.4a to exist within project plans and for a package based on this simplified concept to be useful; experience shows, however, that this is not generally the case. It is necessary therefore for a computer-based planning package, if it is to use hierarchical structures, to provide for and recognise cross links.

It has been shown above that although cross links may be inevitable in all but the most simple hierarchical plans, the cross links can with confidence be predicted to be concentrated in two areas: those activities which form part of

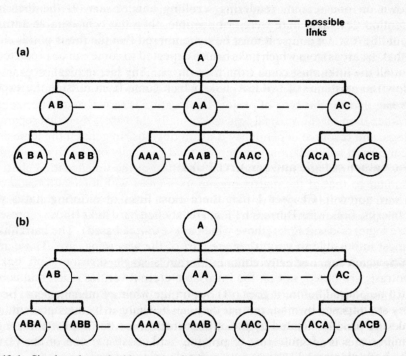

**Figure 13.4** Simple and complex hierarchies; (a) without cross linkage; and (b) with cross linkage.

the same division as the activity being considered, and those activities which share the same resources.

These limitations of the areas of likely cross linkage are important if the benefit of the simplifying effect of hierarchical structures is to be gained. The previous chapter showed that hierarchies which are implicit in the way projects are organised and run can be exploited so as to enable the designer of the project activity to concentrate his attention on particular parts of the network. It is clear that this reason for using hierarchies disappears if the areas of likely cross linkage are ill-defined or if the sources of linkage cannot be predicted. The limitations discussed here thus provide important simplifications enabling the potential of hierarchies to be realised.

The benefit of these limitations becomes even more apparent when graphical means of data input are being developed for use as part of the analysis system. Chapter 4 of this book showed that graphical input is both pleasant to use and efficient. Its use becomes most efficient if the amount of scrolling of the display is reduced to a minimum, that is that the amount of material to be displayed as support for a given decision is minimised. This is especially the case for the linkage matrix, where horizontal scrolling is often needed between each data insertion for large networks.

The hierarchy, reinforced by the ideas explained here (i.e. by enabling the limitation of the display to two well-defined areas), makes possible this limitation of the displayed material so that often all the information can be shown on one screen, rendering scrolling unnecessary. If the focusing of attention described here were not possible, then this considerable advantage would be lost. Of course it must be remembered that the thesis of this chapter is that the areas from which links can be expected to come can be predicted, not that all the links must come from these areas. Any hierarchical program must allow the existence of 'wildcat' links which come from outside the expected areas.

## The nature of logic and resource links

It was noted in Chapter 1 that there exist links of differing status within networks. Following Burmann these are labelled hard links (those representing pure logic) and soft links (those which are resource based). The hard links are almost impossible to vary; they are part of the way things are. These are the links which are normally confined within a single division. Soft links, by contrast, are based, not on the invariable eternal verities (such that concrete must be poured before it goes off) but on the whim of management. Because they are imposed by management they can be changed by management. These links are usually dictated by policy concerning resource deployment, e.g. the number of sets of shuttering to provide, and constitute one of the variables which the planner will adjust during the planning process.

## SPECIFICATION OF HIERARCHIES

It is unusual for these differing types of link to be differentiated during the planning process or in the subsequent use of the plan produced by it. This confusion of the link types is dangerous, particularly where the user of the plan is not the person who formulated it, for soft links come to be regarded as hard links and the manager loses some of the flexibility of action which should be his and which, as has been argued in this book, it is important to retain if the plan is to be effectively used during the whole of the period of the project. Thus the focusing separately on first hard links and then on soft links which has been suggested above helps to separate within the analysis what should be kept separate throughout the whole of the planning/control process.

The separation of these links within the analysis can be carried forward through the whole process to the presentation of the network. Figure 4.3 showed how the use of colour could improve effectiveness by highlighting the critical path in a network. Similar plotting routines can, using either colour or dotted line, differentiate between the hard and the soft links and so help the manager as he seeks throughout the project to avail himself of the flexibility of control.

### The building of hierarchies

As has become clear, the majority of links for a particular activity can be defined using quite small linkage matrices. It seems that two such matrices should be used for each activity, for each activity exists at the intersection of two almost autonomous sets, those of association by logic and association by resource. Of the two sets the logic set (divisional) must always be primary, for the network it implies is not usually variable. The resource set, on the other hand, describes a network which is in the control of the manager and is thus a design variable.

A usable design method is to define divisional links in the way suggested for unilevel networks proceeding down the hierarchy until sufficient detail is achieved. Logical cross links may be present at this stage, but if their number becomes excessive it is likely that the original choice of divisional grouping has not been wisely made. At this stage the project network can be analysed and resource histograms can be produced. Using the resource histograms the planner can decide which resources are to be used as the basis of soft links and he can then ask the computer to display a linkage matrix with just the activities, wherever they occur, which use the specified resource. This adjustment of the resource networks will form the majority of the design iterations.

### A practical approach to the specification of hierarchies

This chapter has described ways in which the simplifying capabilities of hierarchical networks can be used without the stultifying restrictions which are inevitable if cross linkage is prohibited. The identification of the source of the majority of cross links as resource constraints has resulted in a design method which enables the planner to handle complexity, to maintain a flexible approach throughout the life of the project and yet accurately to model the project.

# 14
# THE STORAGE OF DATA FOR HIERARCHICAL PLANNING

The previous two chapters have shown how hierarchies are essential for the handling of complex decisions and how the structure of the networks used within project planning is such that the theoretically very wide possibilities of linkage are in practice limited to a large extent to two well-defined areas of influence. This chapter and the next will seek to build upon the ideas of the previous chapters and develop a method of representing planning and control hierarchies in the computer. First a method of storing data for hierarchical programs will be discussed and in the next chapter a usable computer program will be described.

It was shown in Chapter 13 that, as far as possible, a divisional type of organisational structure would divide the work such that as few as possible logic-based links crossed the divisional boundaries. If this type of organisation is carried over to the setting up of the sections comprising the intermediate stage of hierarchical networks, and if the ideal of autonomous divisions is achieved, then a two-level network will be as shown in Figure 14.1. Here four sections are shown to be independent at the top level, but of course, a logic independence between sections at any level is not necessarily always present; indeed, experience suggests that logical interdependence between sections may be the exception rather than the rule.

It would be unusual indeed to find a project which could be as neatly divided as that shown in Figure 14.1. The previous chapter showed that cross links between sections are likely but that, if the divisions are well selected, it is unlikely that many of them are logic links; it is likely that any cross links between the sections of a divisional organisation represent the constraints on the manager due to the resource limitations imposed on the network by him. The resource constraints form networks which are completely independent of the logic network. Two such possible networks for the logic network shown in Figure 14.1 are shown in Figure 14.2.

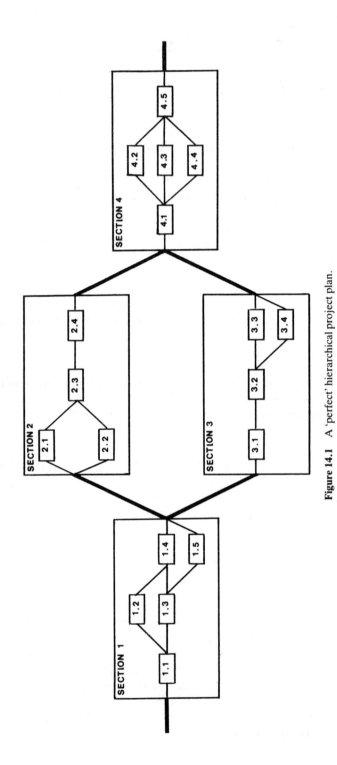

**Figure 14.1** A 'perfect' hierarchical project plan.

# COMPUTER REPRESENTATION AND ANALYSIS

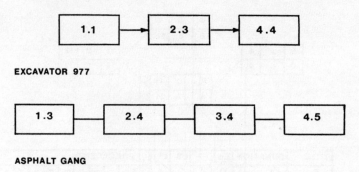

Figure 14.2  Resource-based networks.

Although these networks are themselves very simple, and it should be remembered that the constituent networks are those simple diagrams contained in Figure 14.2 and the individual sections of 14.1, their combination is quite complex. Figure 14.3 represents the diagram drawn without the simplifying help of the hierarchy. It is clear that in project planning as in all other fields of design, the hierarchical structure of complexity can be used to simplify the decision process.

## The computer representation and analysis of a hierarchical structure

Although hierarchical networks can conveniently be shown graphically as in Figure 14.1, their representation within a computer program is not quite so easy. Networks are fundamentally one-dimensional and each activity can be stored such that it carries details of the location of the data referring to its precedents. This is the system which has successfully been used in the programs

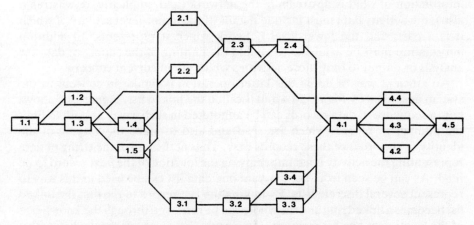

Figure 14.3  The combined logic and resource network.

## THE STORAGE OF DATA

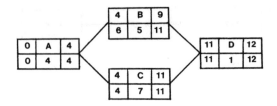

| TITLE | DURATION | ES | EF | LS | LF | PRECEDENTS |   |   |   |   |   |   |   |   |   |
|-------|----------|----|----|----|----|------------|---|---|---|---|---|---|---|---|---|
| A | 4 | 0 | 4 | 0 | 4 | - | - | - | - | - | - | - | - | - | - |
| B | 5 | 4 | 9 | 6 | 11 | 1 | - | - | - | - | - | - | - | - | - |
| C | 7 | 4 | 11 | 4 | 11 | 1 | - | - | - | - | - | - | - | - | - |
| D | 1 | 11 | 12 | 11 | 12 | 2 | 3 | - | - | - | - | - | - | - | - |

**Figure 14.4** The data form for a single-level network.

presented in the earlier chapters of this book and which has the major advantage of simplicity. Because the data addresses contained in the activity data can be changed (by, for example, the addition of activities which precede the listed precedents) the data store must be dynamic, but this causes no serious problems.

Figure 14.4 shows how a simple network is stored in this one-dimensional form, the precedent numbers being the address of the data record representing the precedent activity. For a single-level plan data records such as those in Figure 14.4 are read sequentially into the program, because the whole of the data is required each time an analysis is carried out. The analysis module uses the precedent numbers to guide it to the desired data location.

A hierarchical network has an extra dimension, for, in addition to carrying information of what is upstream in the network (and, implicitly, downstream also), the activity data must include details of the upper-level activity of which it is a part, and the lower-level network which it represents. In addition information must be available to enable the limiting of the supply of data for analysis or output to only those activities which are of current concern.

An efficient way of doing this latter operation is similar in concept to the system described by Figure 14.4 and is called the linked list. Figure 14.5 shows how a sequence of data records can be annotated in such a way as both to limit the number of records which are read and also to make the varying of the identity or the order of these records easy. Thus at the end of the string of data representing the activity is a number giving the location of the next record to be read. As can be seen from the diagram one data list can be used in this way to represent several discrete lists. By linking the last record to the first the linked list becomes a linked ring and the whole can be retrieved through the knowledge of the location of one list member. Here again, however, we are dealing with a

# COMPUTER REPRESENTATION AND ANALYSIS

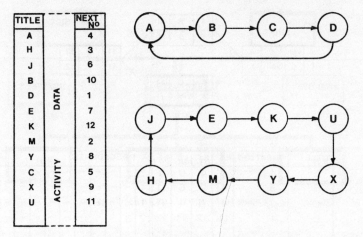

Figure 14.5  Data storage in linked lists.

one-dimensional structure, although now one with the possibility of containing several discrete lists within one large data store and of gaining separate access to them.

In the case of the hierarchical network we must have, within the ring, information concerning the identification of the ring of which the present ring forms a part and also of the rings which make up each of the elements of the present ring. This is shown diagrammatically in Figure 14.6. Each element in the intermediate ring in the figure must carry the identification of the upper level ring and also of the lower level ring which gives details of its composition. It should be noted that, because linked rings rather than linked lists have been used, and therefore the identification of one element in a ring is sufficient for the identification of the whole, each element at the intermediate level ring

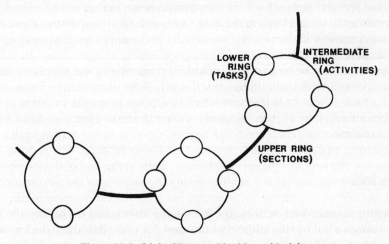

Figure 14.6  Linked lists used for hierarchical data.

## THE STORAGE OF DATA

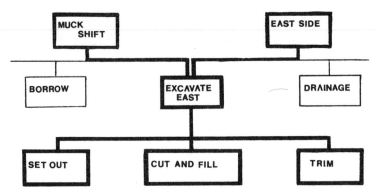

**Figure 14.7** Intersecting hierarchies.

carries the same upper-level identifier and each element at the lower level is defined by just one representative identifier.

The use of linked rings as the basis for hierarchical structures is thus a convenient way of storing both the structure and the element data within one data record. It permits the fast retrieval of data and so brings the benefits of high speed, so important for the interactive approach, to those machines which are limited in power. Linked rings have one grave disadvantage, however, for while they are convenient and elegant in specifying simple hierarchical structures, they are less elegant when there is a possibility that the same element may occur in separate rings of the hierarchy.

It has been argued previously that cross linkage between sections may be minimised by the way in which the project is divided. In a simple project it may be possible to avoid cross links completely if only 'hard' links (logic links) are included in the network. If soft links (resource links) are also included, then this simple solution no longer exists and one element may be present in several different hierarchical paths. This situation is shown in Figure 14.7. The activity 'excavate east' occurs both in the ring representing the earthworks and also in that representing the east side. As argued previously this activity thus is the intersection of two activity sets.

This situation can be handled by linked rings by the use of more than one upward identifier in the data list, but if this is done the structure loses some of the elegance which made it attractive and the programmer is forced to limit the number of calls which can be made to a particular element. This may be inconvenient.

**Index lists**

If the programmer wishes to escape from this rather inelegant solution, then his only recourse is to abandon the attempt to combine data and structure information and to store the information concerning the structure in a separate

## INDEX LISTS

| Name | Precedent | | | | | Index | | Lists | | | | | | |
|---|---|---|---|---|---|---|---|---|---|---|---|---|---|---|
| 1   | 0  | 0  | 0 | 0  | 0 | 1  | 2  | 2  | 3 | 3  | 4  | 4 | 5 | |
| 2   | 1  | 0  | 0 | 0  | 0 | 5  | 0  | 6  | 0 | 7  | 0  | 8 | 0 | 9 | 0 |
| 3   | 1  | 0  | 0 | 0  | 0 | 10 | 0  | 11 | 0 | 12 | 0  | 13 | 0 | |
| 4   | 2  | 3  | 0 | 0  | 0 | 14 | 0  | 15 | 0 | 16 | 0  | 17 | 0 | |
| 1.1 | 0  | 0  | 0 | 0  | 0 | 18 | 0  | 19 | 0 | 20 | 0  | 21 | 0 | 22 | 0 |
| 1.2 | 5  | 0  | 0 | 0  | 0 | | | | | | | | | |
| 1.3 | 5  | 0  | 0 | 0  | 0 | | | | | | | | | |
| 1.4 | 6  | 7  | 0 | 0  | 0 | | | | | | | | | |
| 1.5 | 7  | 0  | 0 | 0  | 0 | | | | | | | | | |
| 2.1 | 0  | 0  | 0 | 0  | 0 | | | | | | | | | |
| 2.2 | 0  | 0  | 0 | 0  | 0 | | | | | | | | | |
| 2.3 | 10 | 11 | 0 | 5  | 0 | | | | | | | | | |
| 2.4 | 12 | 0  | 0 | 7  | 0 | | | | | | | | | |
| 3.1 | 0  | 0  | 0 | 0  | 0 | | | | | | | | | |
| 3.2 | 14 | 0  | 0 | 0  | 0 | | | | | | | | | |
| 3.3 | 15 | 0  | 0 | 0  | 0 | | | | | | | | | |
| 3.4 | 15 | 0  | 0 | 13 | 0 | | | | | | | | | |
| 4.1 | 0  | 0  | 0 | 0  | 0 | | | | | | | | | |
| 4.2 | 18 | 0  | 0 | 0  | 0 | | | | | | | | | |
| 4.3 | 18 | 0  | 0 | 0  | 0 | | | | | | | | | |
| 4.4 | 18 | 0  | 0 | 12 | 0 | | | | | | | | | |
| 4.5 | 19 | 20 | 21| 17 | 0 | | | | | | | | | |

— Index Location
— Activity Data Location

**Figure 14.8** Data storage through index lists.

index. Although such a system may seem to be unattractive because it increases the number of records (and perhaps files) which must be read, it brings with it the simplicity which is present in linked lists applied to simple hierarchies but which has been lost. At its simplest such an index consists of a list of locations of the data records stored in the order in which they are required, but in a hierarchical structure this one-dimensional form must be augmented to include the upward and downward identifiers. Figure 14.8 shows how such a two-dimensional index could be used to store the details of the hierarchical structure of the network shown in Figure 14.1. It is clear that there is now no difficulty in multiple references to a single data element and that if such a reference is made there is no ambiguity concerning the identity of the upper level.

The index lists suggested here have been used in the program to be described in the next chapter, they are a utilitarian solution to the problem of data storage but, unlike some other utilitarian philosophies, have the advantage of being useful.

# 15
# A HIERARCHICAL PROGRAM USING INTERACTIVE GRAPHICS

The previous chapters have shown that hierarchies are a useful way of reducing the complexity inherent in project planning. In this chapter a computer program will be described which can form the basis of a hierarchically structured planning package and which exploits some of the insights gained in the previous chapters and in the development of the programs described earlier in this book.

Clearly the sophistications which have been shown in Chapters 7 and 11 could have been included in the program described here, but they have been omitted in order to enable concentration on the hierarchical aspects of the program.

The purpose of the program described here is to demonstrate the ease with which very complex networks can be built and serviced if a hierarchical structure is used together with a high-quality graphical display. The program has been written in Pascal using an ICL PERQ.

### The use of graphics

In common with the other programs which have been described in this book a major concern in the writing of this program has been to exploit to the maximum the graphical capability of the available hardware. Again this has resulted in a program which minimises the amount of typing necessary for the operator through the use of cursor-accessed menus and the presentation of output solely in graphical form. Where possible the output itself has been used for program direction.

Figure 15.1 shows the form of the program options. As in the other programs described in this work the operator is led through a series of menus. The menus

# THE USE OF GRAPHICS

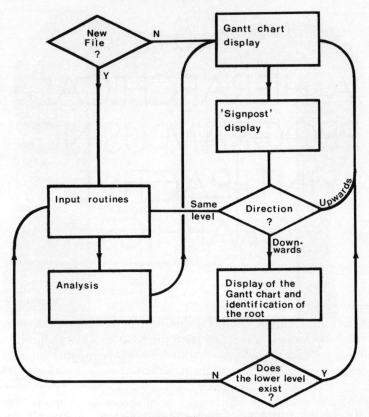

**Figure 15.1** The menu structure.

used are similar in form to those used previously, again the emphasis being on providing the operator with a system which needs little tuition. A hierarchy, of course, contains a new dimension of movement, and thus a new menu is required for the direction of the program through the hierarchy as shown in Figure 15.2. If the operator wishes to proceed upwards, then the program automatically generates the Gantt chart for the upper level or exits from the program; if he wishes to proceed at the same level, then he is given the data entry menu; if he wishes to descend, then he is shown the current Gantt chart and asked to indicate which activity he wishes to expand.

The output for the program can be simpler than that required for the unilevel network in that it is unlikely that scrolling will be normally necessary. In this particular program the number of elements at each point in the hierarchy has been limited to twenty. This seems to be a reasonable number if the simplifying power of the hierarchy is to be realised, and it renders vertical scrolling of the screen unnecessary. Of course scrolling of the type described previously could be incorporated into this program but at the risk that some of the potential of the hierarchy to simplify the building of the network would be lost through the use of very wide spans within the hierarchical structure.

# HIERARCHICAL PROGRAM USING INTERACTIVE GRAPHICS

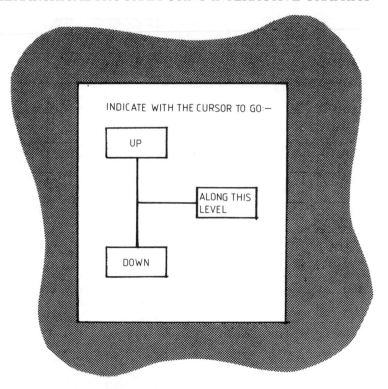

**Figure 15.2** The 'signpost' menu.

As in the earlier programs Gantt charts are used exclusively for data presentation. The facility to interrogate the Gantt chart has been omitted purely to avoid complicating this development program; it should be reintroduced into a production version.

Broadly the same data input techniques are used as were successfully developed previously. The dependency matrix within a hierarchical program can, however, be simpler than that required for a single-level program because of the removal of the need for complex links.

The complex links in a single-level network reflect the implied presence of a lower-level network. For 'A must not start until one day after the start of B' implies that within B there is a task upon the finish of which depends the start of the first task contained in A, that is that both A and B themselves consist of networks. It is clear that in a hierarchical network that which is implicit in a single level network is made explicit, and therefore that the need for such devices as complex links disappears.

## The specification of resources

The need to identify the elements of the network (at any level of the hierarchy) which are related by resource need makes the specification of resources, which

## THE SPECIFICATION OF RESOURCES

has been seen as an optional extra in the programs described earlier, absolutely essential.

In order to help the user quickly to find the resource he needs amongst what could be a very comprehensive resource list, use can be made of the inherent grouping of resources.

If a user of the program wishes to specify a particular resource for an activity he is first asked to identify the resource group by means of the cursor and a graphical display of the available groups. Each of the resource groups can contain up to ten resource teams, and the indication of the group enables the computer to display a grid as shown in Figure 15.3. Using this grid the operator can indicate resource need in the same way that logical links are indicated on the dependency grid; a dependency grid containing only those activities which share a common resource can then be generated to allow the specification of 'soft' links.

The use of graphics for direction and presentation seems to be essential to the success of hierarchically structured programs. In the absence of graphics the direction of the program and the assimilation of output would be slow and, because of the complexity of the structure, difficult to use. Although the general adoption of hierarchically structured networks may be considerably in the future, the work on graphics reported here suggests that this will be a fruitful approach which will be necessary when such networks eventually appear.

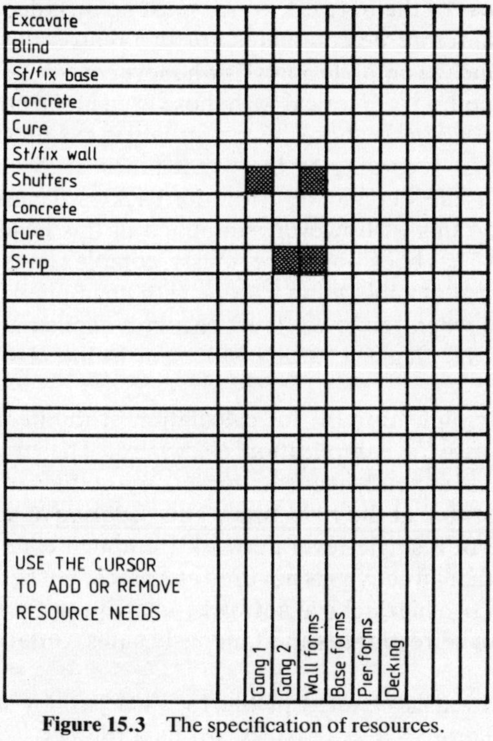

**Figure 15.3** The specification of resources.

## The analysis of hierarchical networks

One of the advantages which flows from the restricting of hierarchical structures to those which are simple (that is those which do not contain cross links) is the reduction in analysis effort which becomes possible. A simple hierarchy can be analysed completely by considering only those elements which have suffered change and realising that the identity of those elements can be restricted to those which lie on the path through the hierarchy to the new or altered activity.

Unfortunately, although this restriction of effort is theoretically possible, it has been shown in the previous chapter and, more importantly perhaps, by experience, that simple hierarchies of this type rarely exist in the planning of projects. In the complex world of project management cross linkage is inevitable unless either the project is so simple as to make the use of multilevel programs questionable or each division is set up so as to be functionally autonomous. This autonomy may occur between sections of large civil engineering projects, but breaks down as a descent is made through the hierarchy to more detail where the sharing of resources is inevitable and, because of the need to maximise resource use, highly desirable. In small projects this sharing occurs at all levels.

This inevitability of the presence of cross links within a hierarchy prevents the construction planner from simplifying the analysis, as he may in other fields of management, for if cross links are permissible, then it is not possible to predict those areas of the network which must be included in the analysis. All analyses must therefore be complete. For this reason the algorithm must be designed to produce a complete analysis whenever or from wherever it is called.

The analysis method is identical with those used in the other programs which have been described in this book. Each activity is examined in turn and event times are adjusted according to the event times of the precedents (for the forward pass) or the dependents (for the backward pass). The analysis of hierarchical programs is, however, simplified at this level by the absence of complex links. As has been noted previously, complex links are introduced into unilevel programs to enable them to cope with an implied lower level, but this need is removed when the lower level becomes explicit and thus the analysis method regains the elegance which is present in the algorithm illustrated in Figure 4.1.

Clearly the simplification of this calculation at the heart of the analysis is bought at the price of complication elsewhere. The process of analysis as described above is dependent for its success upon the various activities being fed into the algorithm at the right time. This ordering of activities is relatively easy in the case of a single-level network (although care must even here be taken to ensure that an activity is not used as a precedent before it has itself been operated upon). In a hierarchical network, where progress through the data list proceeds in two directions simultaneously, this ordering becomes more difficult.

This complication is handled in the forward pass by using the algorithm illustrated by Figure 15.4. The backward pass follows a similar form. In this

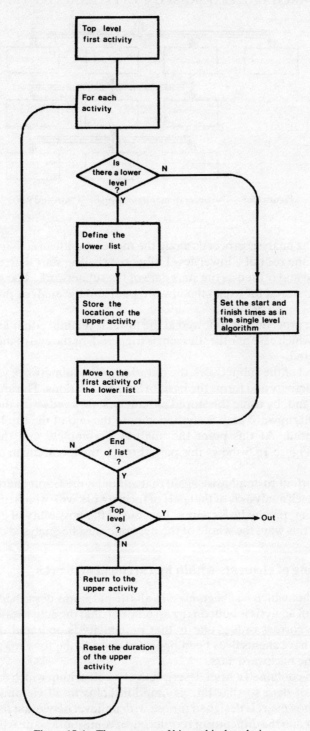

**Figure 15.4** The structure of hierarchical analysis.

# HIERARCHICAL PROGRAM USING INTERACTIVE GRAPHICS

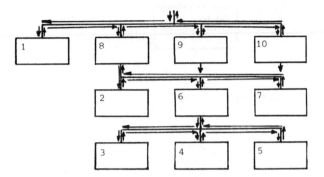

**Figure 15.5** The order of hierarchical analysis (forward pass).

algorithm the analysis proceeds along the top level until an activity is flagged as being itself the root of a lower-level subnetwork. The start date of the activity is calculated and is used as the start date of the subnetwork. The parameters of the root are stored for use in the upward path and the analysis proceeds to the lower level.

The same procedure is followed at the lower level until either another root is flagged, in which case another descent is triggered, or the end of the subnetwork is encountered.

At the end of the subnetwork the duration of the subnetwork is carried back to the root activity and forms the basis of later calculations. The analysis returns to the root and, by using the stored parameters, proceeds with the upper-level analysis. This upward path is continued until the end of the top-level network is encountered. At this point the analysis is complete and the results are displayed. Figure 15.5 shows this path through the hierarchy in diagrammatic form.

It is important to emphasise again that although the Gantt chart produced at the end of each analysis is of that part of the network with which the planner has been working, the analysis cannot, because of the possibility of cross links, be confined in this way; the whole of the network must be analysed each time.

## The ordering of elements within hierarchical networks

The algorithm which has been used in all the programs described in this book as the basis of analysis is built on the assumption that the activities are presented to it in the correct order, that is that no activity is operated upon until its precedents have themselves been operated upon in the forward pass, and the reverse in the backward pass.

Chapter 5 contained a brief description of an algorithm which carries out the reordering of data so that this assumption holds in all circumstances. This algorithm, however, is designed for use with unilevel networks; the problem of wrong order and the difficulty of reordering become more acute when hierarchi-

# THE ORDERING OF ELEMENTS

cal structures are considered. In hierarchical networks the links to an activity may be shown on several link matrices and may involve links which cross hierarchy paths, in this circumstance the elegant solution developed in the earlier chapter is inadequate and an alternative must be sought.

In spite of this difficulty it is clearly prudent to use the simple unilevel algorithm as far as possible to 'clean up' the data presented for analysis at each hierarchy level; this process is incorporated immediately after each call to the routine for the input of local logical links and ensures that the 'wrong order' activities are not carried forward into the analysis routine or to the data store. The major difficulty, however, remains, for links within a hierarchy are likely to come from any part of the structure, largely from resource links but also from the few logic links which cross the hierarchy boundaries.

These cross links can be random in origin and destination and are therefore very difficult to reorder in a logical and rigorous fashion using the type of algorithm described above. The source of the difficulty is that more than one cross link can exist between two hierarchy paths and these cross links can be in opposite directions, making the reordering of hierarchy paths impossible.

In view of this impossibility an alternative approach must be sought, and an iterative method can be successfully used. Figure 15.6 shows a two-level network which has cross links in opposite directions between two paths. It is clear that if only link 1 were present the network could be solved by reordering A and B, but this solution is made impossible by link 2 which requires that the two root activities should remain as they are. The two low-level cross links imply that links between A and B are flowing simultaneously in both directions.

The solution to this problem is to repeat the analysis of the network until all the event times of all the activities are stable. This process is shown in Figure 15.7, where the early start figures for each activity are noted; as can be seen, the

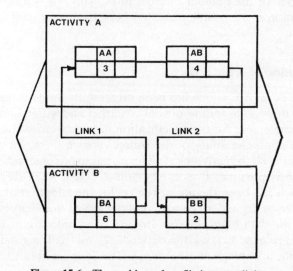

**Figure 15.6** The problem of conflicting cross links.

## HIERARCHICAL PROGRAM USING INTERACTIVE GRAPHICS

| ITERATION | ACTIVITY | DUR | EARLY START | EARLY FINISH | LATE START | LATE FINISH |
|---|---|---|---|---|---|---|
| 1 | A A | 3 | 0 | 3 | 0 | 3 |
|   | A B | 4 | 3 | 7 | 3 | 7 |
|   | B A | 6 | 0 | 6 | 3 | 9 |
|   | B B | 2 | 7 | 9 | 7 | 9 |
| 2 | A A | 3 | 6 | 9 | 6 | 9 |
|   | A B | 4 | 9 | 13 | 9 | 13 |
|   | B A | 6 | 0 | 6 | 6 | 12 |
|   | B B | 2 | 13 | 15 | 13 | 15 |
| 3 | A A | 3 | 6 | 9 | 6 | 9 |
|   | A B | 4 | 9 | 13 | 9 | 13 |
|   | B A | 6 | 0 | 6 | 0 | 6 |
|   | B B | 2 | 13 | 15 | 13 | 15 |
| Subsequent iterations give the same result | | | | | | |

**Figure 15.7**  The growth of accuracy during iteration.

repetition of the analysis eventually captures the errant cross links. If a logic loop is present, then stability will never be reached.

Analysis of this method shows that the maximum number of iteration loops required is equal to the number of cross links but that this maximum would be required only in the most extraordinary circumstances (if all the cross links were running in directions counter to logic and if they were nested so as to create the maximum mutual interference). In practice the number of iterations is less than 25% of the number of cross links. This is a workable if less than beautiful solution to the problem and has been incorporated into the program described here.

### The specification of cross links

The dependency matrix, which has been successfully used for the specification of links in the unilevel programs described earlier and within each hierarchical layer, is not convenient for the specification of logical cross links. The reason for this is the unpredictability of the source of such links, for a dependency matrix containing all the activities at all levels has to be constructed. While such a construction is possible, it is cumbersome, and it loses the advantage of simplicity which has been the major reason for the adoption of a hierarchical structure. It can be argued, of course, that the loss of simplicity is confined to a small part of the data input, because logical cross links are an exception and certainly not a rule, so that a large dependency matrix is a good solution as it brings with it the advantage of consistency. The designer of future systems must decide this question for himself.

# HIERARCHICAL NETWORKS – POTENTIAL

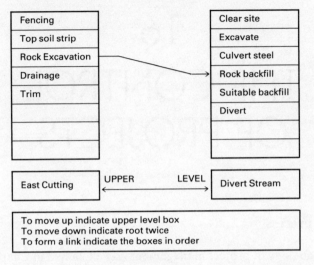

Figure 15.8  The graphical specification of logical cross links.

An alternative to the dependency matrix can be used, a system of specifying links which has been found to be suitable also for the unsophisticated graphics of microcomputers. Two lists, one containing the dependents and one containing the precedents, are displayed side by side. A link is established between precedent and dependent by indicating first one and then the other in their respective lists, and a line drawn on the screen to represent the new link. This display is illustrated in Figure 15.8. In a unilevel network both lists must be scrollable, but scrolling should not be necessary in a hierarchical program. In this case the dependency list contains only those activities in the current hierarchy element and the precedent list can be changed by proceeding up or down the hierarchical structure until the required element is located.

Cross links are a necessary but messy addition to hierarchies; the routine described reduces the mess but does not take it away completely.

## Hierarchical networks – potential

This is not the first network program to incorporate subnetworks, but its approach to cross linkage may make it more useful than most. Previous programs have been little used because the nettle of cross links was not grasped, and it is clear that unless this problem is solved the hierarchical program is of little more than passing interest. Two factors suggest that the future importance of hierarchical-based programs may be considerable: the first is the combination of graphics with the logical recognition of the source of cross links due to resources, which is developed here and considerably reduces the tedium of data entry; the second is the way in which the hierarchical structure mirrors the information needs of management during control.

# 16
# THE CONTROL OF PROJECTS

**The control process**

The journey analogy of project management was explained in Chapter 1. This analogy likens the project to a journey and the manager to a navigator and, as was noted in the earlier chapter, its aptness to the control process is shown by the use of the term 'cybernetics' for the science of control.

It was suggested, using the map analogy, that the planning of the route of the journey could not be rigidly done before the start – the planning must expect and allow for variations made in the light of the experience gained on the journey itself. Applying this insight to the control of projects, it became clear that there can be no hard boundary between the planning and control functions of project management; plans formed early in the design stage should remain fluid until the work to which they refer is carried out. Planning and control are thus one design process carried out in the light of increasingly accurate and detailed data. Viewing control in this way causes one to question the wisdom of an organisational structure which places planning and control under the control of separate sections within the organisation. The journey analogy thus provides valuable insight into the way in which plans are formulated and used.

A second way of viewing project control is to see it as a species of the control which is exercised automatically on machinery. Watt's steam governor is the example of mechanical control which is normally quoted in management texts. This device is illustrated in Figure 16.1.

The steam governor relies, for its stability, on an almost instantaneous response. Steam pressure rises, the speed of the engine increases, the brass balls are driven outwards by centrifugal force, the steam pressure is reduced. If the response of the engine to the rise in steam pressure, or of the valve to the engine speed, is sluggish, then the valve may act in an inappropriate way and an unstable speed will result.

It is clear that the flow chart in Figure 16.1 representing the action of the governor is similar to the diagram of a project control system shown in Chapter 2, and that in principle the systems are identical. There is one important difference, however – that of response time.

## FEEDFORWARD CONTROL SYSTEMS

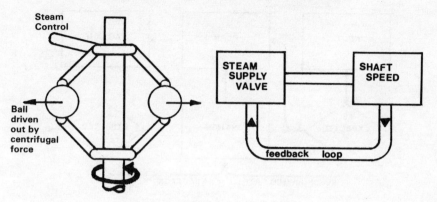

**Figure 16.1** Watt's steam governor.

The map analogy shows clearly the difference between the two control devices. The captain on the bridge of the ship described in the analogy must become aware of the changed circumstances, must decide on appropriate action, and must convey his instructions to the engine room staff. All this takes a finite time during which the ship is ploughing along on the previously set route. The steam governor, on the other hand, imposes the sort of control that railway lines impose on a locomotive, instant correction and return to a pre-ordained route. Project control is much more like the ship than the locomotive because the route cannot be known in great detail, because the steering mechanism requires a long response time, and because the collection and digestion of data are lengthy processes.

## Feedforward control systems

It was suggested in Chapter 1 that one way of overcoming the difficulty caused by the long response time of management control systems was to provide what Koontz and O'Donnell in their book refer to as 'feedforward'. The input parameters to a system rather than the output from the system are measured in a feedforward system of control; control is then exercised in response to them, ahead of the action. Figure 16.2 illustrates this in flow chart form. If such a system was to be applied to a steam engine as a replacement for the steam governor, the incoming steam pressure would be measured directly and the valve activated in accordance with the result of the measurement rather than with the resulting engine speed. While such a system can be valuable for many control purposes, the seeds of its fatal fault can be seen in the application to the steam engine, for engine speed can fall for a number of reasons, steam pressure being one and engine load another. The beauty of the feedback system is that it operates whatever the reason for the change in engine speed, whereas the feedforward system, in the simple form described, is sensitive to only one cause.

# THE CONTROL OF PROJECTS

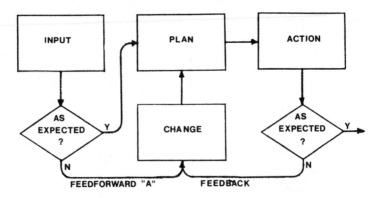

**Figure 16.2** Models of control.

Feedforward systems of control are likely to be most valuable when the input parameters can be carefully specified and are the same throughout the control period. The steam engine is one such system and so is the management of continuous processes.

It is difficult to see how feedforward systems could be used in project management where the input parameters and the activities to which they refer change constantly throughout the period of the project. At one stage the critical parameter may be the delivery of steel, at another the acquisition of certain plant. Feedforward in an informal sense is obviously possible, but a formalised system would be difficult to organise and run.

It seems, therefore, that although feedforward is theoretically a sound concept it breaks down when the process being controlled is not continuous, and that the reason for this breakdown is the difficulty of building an automatic feedforward link which is flexible enough to respond to different parameters as the project proceeds. A working feedforward system must either be so comprehensive that it can respond to the changing project environment, or must abandon the attempt to make the link (link A in Fig. 16.2) automatic and to rely on a different and more informal type of reporting system.

## Short-term forecasting

Recent work by Trimble and others at Loughborough University suggests how this informal replacement of link A might be done. The 'inputs' to the project environment, that is delivery dates, weather forecasts, observations of past productivity, etc., are used just as in the feedforward system described above. In this case, however, no attempt is made to link input data to performance forecasts through an automatic system. The automatic link is replaced by a link which is purely in the minds of those who are reporting the input parameters; these people use the input data to produce forecasts of the short-term future, perhaps the completion date of the tasks current to the project. Thus short-term forecasting replaces feedforward but is identical in concept and effect – the control system can register problems before they occur and corrective action can be taken.

This change in the philosophy of control from feedback to feedforward suggests that a change in the model of the process may be appropriate. The model of Watt's governor is helpful but limited, for the breaking of the link A makes the diagram less than completely appropriate. Short-term forecasting is perhaps best represented by returning to the picture of control as continuing design. As he makes his short-term forecast the manager is feeding into the design process data which were not originally available, and is thus facilitating decisions about the detail of the design which, as has been argued, are wisely left until late in the project. This flow of data can conveniently be represented by a variation on the diagram which was quoted in the first chapter. This shows the decreasing freedom of design as the project decisions are taken and acted upon and the data available to the designer become increasingly complete.

If control is viewed in this way, as the provision of increasingly complete design data, then some doubt must be cast on those systems which yield only data concerning past performance, for such data are of only indirect relevance to the continuing design of the incomplete work. The control system must, if it is seen as design, give priority to two features: a flexibility of approach which allows continuing design, and facilities which allow the manager quickly and conveniently to note his predictions of the future. Before looking at the implications of this approach to the design of computer-aided systems it is useful to look at the control systems at present being operated within the construction industry.

## Construction control systems at present employed

In the non-computer environment in which many of the smaller companies in the UK construction industry operate, the first of the two requirements noted above, that of allowing flexibility of plans and planning, has largely been abandoned. The working project manager simply does not have the time formally to replan the project whenever changes in constraints occur. He will normally confine what changes he does make to those which can be accommodated within the original plan. It may be that not only is he constrained by time but also by organisation; many large organisations have planning departments which effectively take responsibility for any replanning from the hands of the project manager, thus introducing both a time lag and a boundary over which information must flow.

The control system on a typical construction site, if it exists in other than name alone, usually has a response time which renders the data produced of little value for the design purpose being discussed. The data may be invaluable for the preparation of plans and estimates for future projects, and a good system would ensure that the data were fed back to those in head office who need it for these purposes, and it will be of help to the company's accountants when they are preparing their final accounts for the project, but it is of help to the project manager only if it can help him re-evaluate estimates of productivity, cash flow, and float. Unless the response time of the system is considerably less than the time taken for a typical individual activity on the project, the historical or feedback system of control cannot be of direct use.

This difficulty has been recognised and the introduction of computers into head office organisations has been seen by some as an opportunity to remove it. A central computer can provide a store making the data generated on site available to the planners and estimators; it is possible to provide a menu of rates and productivities for different resources in different situations and thus to provide a (somewhat crude!) automatic estimating system for incoming bills of quantity. The data storage of such a system, properly organised, can thus benefit many within the company.

Although potentially systems such as these can provide at the minimum a fast, flexible, and easily accessible data store (and it must be said that few have even approached this minimum) the provision of a computer based at head office has not improved site control. The data store is retrospective, and thus of limited use to the project manager; the longer route for the information has the effect of increasing the response time of the system, and the form in which the data are presented is often totally inappropriate for the project manager.

The project manager needs information about those activities which are of immediate future concern and which the latest available data suggest may be different in some way from the plan assumptions. What he often gets is blanket data for the whole of the job whether or not similar activities are to be carried out in the future as part of his project and whether or not the activities were performed as planned.

Thus the provision of centralised 'control' can have an adverse rather than a beneficial effect on site management. The manager can be under the false security of a useless control system and so abandon the more informal methods of site control which had previously served him well and which, because they were more local, were richer in information content. In addition to this reduction in the quality of the information, the additional transfer of information (to head office, to punch girl, to computer, to senior management, to site) will lengthen the response time and thus further reduce the utility of the data to the project manager for the task in hand. This extended control loop is shown in Figure 16.3.

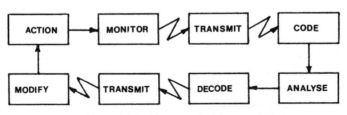

**Figure 16.3** The control cycle extended.

## THE NEED FOR COMPUTERS

This brief analysis shows why, despite head office claims to the contrary, centralised control systems are rarely popular with site management, and why the wise site manager will not rely on such a system for day-to-day control. The centralised system is a useful 'data capture' device and should, in time, improve estimating accuracy, but in its present form it has little to offer at site level.

### The use of short-term forecasting in the industry

The computerisation of control which has been described above relies on the feedback model, on which some doubt has been cast in this book. The object of such computerisation has been to reduce the response time and thus to make the system more effective. But if this failed to achieve this objective and if it brought with it other adverse effects, then perhaps a different approach should be tried. One such is the short-term forecast to which reference has already been made. Short-term forecasts look to the immediate future, making predictions of productivity and cost which will be more accurate than could have been made at the time of the original planning of the job, and do this in the same formal way that the retrospective data of the feedback control system collected its data.

Like all managerial innovations, short-term forecasting is merely a formalisation of what the good manager had always done. In well-managed projects line managers and foremen are asked what difficulties they anticipate in the following short period of time and decisions for future action are made so as to avoid these difficulties if this is possible. If the difficulties are significant in the context of the whole of the project, then strategic decisions are centrally made.

Thus the good manager, by communicating informally with his staff, produces control data which are forward looking and which pick up and emphasise only those project areas which are expected to produce difficulty. The problem of the swamping of the manager with data which do not require his attention is solved by the selection of the data at the point of input to the system, by the foreman or line manager.

Many project managers, moving towards a formal system, ask line management to produce short-term plans prior to regular management meetings, and thus force their staff to make judgements concerning the short- and medium-term performance of their sections and/or functions. The use of these plans is usually restricted to the meeting for which they were produced, but there is no need for this limitation. It is clear that the formalisation of good practice which is emerging here can only be beneficial, and this method of improving site control seems to have value both within and outside the computer environment.

### The need for computers

It is clear from the discussion of non-computer based systems that the computer can hinder the development of useful ideas by petrifying (and hallowing) unsatisfactory management practice. The provision of a centralised and retrospective control system is possibly an example of this. For some this may be a pleasant reinforcement of their opinion that the computer has little to offer the construction industry in terms of control systems, but others will see that this is not true.

The model of control favoured in this chapter is of a continuing design process, and this model places an extra burden on site management. If the site managers do not have the tools to make this tuning of the design to the latest data easy and quick, they will not do it. They do not have the time and only the computer can provide the tools.

A secondary use of site control systems is the collection and storage of data for future use in estimating and planning. This data can be, and is in many companies, collected by manual means, but this collection is slow, haphazard, and expensive. Data retrieval from a manual system is similarly difficult. A computer can transform this data storage and retrieval, converting it into a tool which is fast and friendly.

Implicit in these remarks, of course, is the overriding requirement that any computer-based system must be such that those called upon to operate it, probably very far from being computer enthusiasts, must enjoy its use. For this reason great emphasis in this area of development, as in the development of planning systems, must be placed on input and output techniques.

# 17
# THE INTRODUCTION OF COMPUTER-AIDED CONTROL

**The computer environment**

Whenever new facilities or techniques become available there is a temptation to introduce them into existing systems gradually without questioning the suitability of the system to receive them. There may be good reason for doing this – large investment of time or money before full evaluation is unwise, sudden changes of method are threatening to staff and may upset morale, carefully built skills should not be squandered – but a gradual approach may fail to exploit the full potential of the new technique.

This sub-optimal performance is more likely to occur when the new technique offers facilities which were not previously available than when an improvement to familiar facilities is offered. In the former case the existing system will have evolved precisely so as to minimise the effect of the missing facility, and provision of the facility in the context of the existing system will quite rightly be seen as unnecessary.

It is possible that the use of computers in construction planning and control are examples of this softly softly approach. It was shown in Chapter 2 how the early attempts to introduce computers into planning systems resulted in many cases in an increase in the time of analysis and in the frustration of all concerned. The reason for this was that the computer was being introduced uncritically into a system designed for manual use.

It is only when the old system is demolished and a new system built around the newly available tool that a partnership between the old skills and the new facilities can be established, and the new system used to anything approaching its potential. The introduction of computers into planning was disappointing and similarly the introduction of computers into project control has not been as

## INTRODUCTION OF COMPUTER-AIDED CONTROL

successful as had been hoped. As the previous chapter showed, the introduction of head-office computing power into a unit costing system can result in very long response times and a system which is not useful for construction control.

To avoid this situation it is necessary to strip away current methods and organisations and to look at what is required of the mechanism being considered. Having done this it is possible to rebuild, taking into account the newly available facilities. It is possible, of course, that at the end of the rebuilding stage the system may be identical to that which existed before, and this should be cause for rejoicing (given that the demolition has really been rigorous), for it is not a necessary part of the introduction of new technology to change everything in sight. It is likely, however, that a system built in this way will be different from one which grows from an existing system without a reappraisal phase.

### The requirements of a construction control system

If it is wished to sketch the shape of a control system for construction activity it is necessary first to define the requirements and the available resources. Firstly requirements will be considered.

The purpose of a project control system is to facilitate the successful completion of the project. Success may be defined in many and various ways, but it is likely that overall cost, profit, completion time and public image are among the parameters used for its evaluation. In addition there are secondary purposes, of which the provision of planning and estimating data for future projects is the most obvious.

In attempting to meet these requirements the designer of the project control system must recognise the factors which are pushing the UK industry towards a mechanised system of control, that is, the resources must be considered. The most important of these resource factors is possibly the difficulty of recruiting and keeping high-quality staff for what is a rather mundane task.

The collection and processing of site data has, in the past, suffered from the fact that it has been carried out by personnel who have been bored by the work that they have been given to do. The reduction in cost and increasing availability of computers obviously provides an escape route from this particular problem, for much of the repetitive work can now be handled automatically, but we must be careful not merely to replace one tedious job with another.

The primacy of the current project in the list of requirements emphasises the need to give aid to the controller in his design function. It has been argued previously that this implies either a very fast response (because of the non-continuous nature of construction projects) or a short-term forecasting approach. The relegation of data collection for record and future purposes reinforces the doubts already expressed about the usefulness of some of the currently used systems.

## Site data sources

There is an apocryphal saying that construction sites are primarily generators of data and only produce permanent works as a byproduct of this function. While the construction engineer, in hotly denying this, can point to many spectacular achievements by his industry, he cannot deny that a vast volume of valuable data is produced on site, nor can he deny that much of it is frequently wasted.

At the present time in the UK construction industry data are gathered by various levels and functions of management and for various purposes. The major sources and destinations are discussed below.

### *Site management data*

Data concerning management decisions and their background are collected informally, usually in diaries, and are mainly used for the retrospective reconstruction of controversial events. An obvious use is the formulation or refutation of claims for extra payment. Anyone who has tried to use these data for this purpose will know that they are very variable in quality and therefore unreliable and difficult to abstract and to use.

Some attempts have been made to formalise this system. Twort (1980) mentions such a formalised system which, although intended for supervisory staff (i.e. the resident engineers staff in the usual UK system) could be adapted for use by the contractor. The basis of this system, which has been used by the author and found to improve enormously the standard of the data, is to replace the diary by a loose leaf file having one page for each day and being preprinted with a proforma giving a standard layout and a series of prompts. The staff at all levels complete these forms daily and pass them to their superior for checking at the end of the week. This system quickly fills filing cabinets but provides a very comprehensive retrospective view of the activity of the project.

### *Bonus measure*

Most construction activity supports an incentive bonus scheme whereby the hourly rate of the workers is tied in some way to productivity. The hourly rates are calculated each week for each group of workers and therefore a weekly assessment of the production of the project is required. The weekly bonus measure is carried out by surveyors employed for the purpose and is formally presented and of consistent quality.

This measure is the source of the progress data for many control systems, but is not ideal for the purpose because the data collection is intermittent, the processing time is long, and the surveyors are not members of the management team and thus not in a position to comment on the reasons for unexpected results or to suggest remedies.

## INTRODUCTION OF COMPUTER-AIDED CONTROL

### *Allocation sheets*

Whereas the bonus measure is concerned with the production side of the production/time equation, allocation sheets provide the data concerning the time spent by each group of workers on each part of the work. The purpose of the allocation sheets is purely to provide bonus data and the allocation sheets are filled in by the first level supervisors, the gangers and foremen. These people will normally themselves be recipients of the bonus payments. In view of this pecuniary interest, it is not surprising that allocation sheets are notoriously inaccurate.

Although allocation sheets are difficult to use for control purposes because of their suspected low accuracy, the data they contain are difficult to obtain by alternative means. Construction sites are very large and only the lowest-level supervisors are in a position to report in detail on the distribution of the time of the working gangs. Inadequate though this system is, the lack of an economic alternative will probably guarantee its continuance.

### *Monthly measure*

The basis of the payment for measure and value contracts is a monthly measurement of permanent works completed. The measurement items are those listed in the bill of quantities and the work is carried out by a team of specially employed quantity surveyors. The monthly measure is not a useful control device because the items are frequently large, they refer explicitly only to permanent works, and the frequency of the operation, monthly, is too low for most control purposes.

Each of these sources of data is effective for its own purpose; each measures different parameters or levels of detail, or measures at different intervals, each measurement is done by a different group of people. Control systems have tended to grow up using one of these sources of data and, perhaps wisely, there has been little effort to try to amalgamate them.

Most frequently bonus data and allocation sheets have been used, often monthly measurement has been used as a coarse indicator of progress. The information from the site diaries of the management team has not been used, doubtless because of its extreme variability and because of the difficulty of its collection and codification.

### **Data capture in the computer environment**

The introduction of computers into the site environment brings new possibilities for data capture, for although site staff are unaccustomed to computer hardware, good input and output routines can breach the wall of jargon which frequently fence the machine in. If such routines are available, then this wall

## DATA CAPTURE

can be broken and the computer input point can be accessible and friendly, so that there is no reason to limit the use of the computer to specialised staff.

If this approach is followed, the source of the data and the input point can be made to coincide and the tortuous information path shown in Figure 16.3 reduced. As a direct consequence of this the response time of a system is also reduced. There are two major difficulties in the way of an information system which seeks to capture the data at its source. These are the need for accessibility and the need for intelligibility. These two difficulties need to be considered together, for they both concern the investment of money in hardware provision.

Accessibility is dependent on the location of the computer input points, and thus on the number to be supplied and on their robustness. Civil engineering sites are not ideal environments for sensitive electronic devices, and this problem becomes more acute if the computer is to be put not in the site head office but in the section cabins where dust, mud, and wellingtons are more noticeable than air conditioners and 'no smoking' signs! At present the need for accessibility suggests a computer at the lower end of the market – cheap, relatively simple, and robust. Unfortunately this type of machine does not meet the other criterion, that of intelligibility.

The construction industry tends to think graphically. Its major data transfer system is the handing on of charts and drawings concerning the thing to be constructed and the method of its construction. One of the reasons for the failure of computer applications in construction in the past has been the inability of the current computer hardware to imitate this information transfer.

The availability of computer hardware which offers graphical I/O is clearly an important development in the solution of this problem. It has been shown in the earlier part of this book that planning programs using interactive graphical I/O are most effectve tools. The same techniques of data input, perhaps utilising a hierarchical data structure, will similarly be useful for the development of tools for the control phase of the project method design. The difficulty comes with the present cost and relative delicacy of the graphical terminals available, for although the cost of such equipment is falling rapidly, it is still more expensive than would be thought sensible for equipment in a site cabin.

The ideal remains a data input point at the location where the data are generated, an input which will allow use by unsophisticated personnel with minimal training – that is, one which will support the high-quality graphics which have been demonstrated earlier in this work. Any system which falls short of this ideal in hardware terms must include a data flow from the data source to the input point. It seems clear that the longer the chain carrying the data, the longer the response time and the more attenuated the data will be.

Two possibilities exist. A computer input point can be made available at the site office. Here the environment is more conducive to office machinery, and therefore less robust plant can be used. The reporting must be frequent so as to avoid long lead times such as accumulate in conventional systems. If this is to be done the reporting must be done by line management; any other staffing arrangement would be prohibitively expensive. A second possibility is to

## INTRODUCTION OF COMPUTER-AIDED CONTROL

increase the accessibility at the expense of intelligibility and to use simple data recorders, cheap and robust, such as are now available in the form of digitising cassette recorders. The major disadvantage of these devices is the need for numerical codes such as are at present used for allocation sheets. Such systems have been shown to be very error prone.

**An interim approach**

Recognising the need for data input at or near its source and the need for frequent reporting it is clear that line management should be the reporting agents. Given that line management is concerned with much more than data recording, the reporting system must be quick and simple to use but rich in information. Graphics provide an obvious method of meeting these requirements.

Graphics are not, of course, confined to on screen displays; hard copy in the form of charts can easily be produced and used as the basis of a data-handling system. Centrally located hardware, that is hardware in the site head office, can produce charts which, graphical in form, are individual to a section of the project and to the particular time and these charts can form the vehicle of the data transfer.

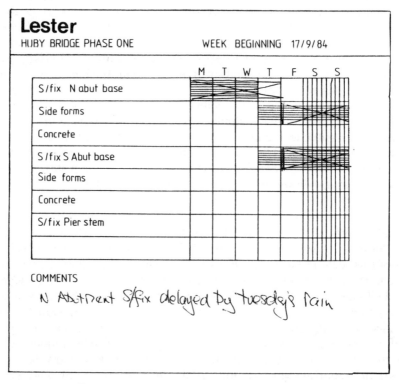

**Figure 17.1** A graphical means of data capture.

## PROVIDING THE FORWARD VIEW

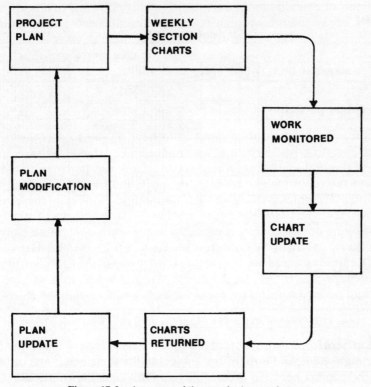

**Figure 17.2** A suggested 'interactive' control system.

A possible system consists of a central computer containing the project model in the form of either a single-level or hierarchical network which is capable of producing hard copy Gantt charts of sections of the job. The charts are limited in extent and so need be no larger than A4 size. They are produced at the start of each week and show the activities to be carried out in that section during the week. Each day the supervisor concerned with the work sketches on to the chart the day's activity, commenting where necessary upon any discrepancy with the plan as presented. The completed chart is returned to the site office at the end of the week and the chart, reproduced on the graphics screen of the computer, is edited in accordance with the information flowing from the site. The plan is updated, new charts are produced for the following week, and the completed charts are stored for future reference. A completed chart is shown in Figure 17.1. The system is shown in Figure 17.2.

## Providing the forward view

Just as the reporting system described above relies on the low-level supervisors, so must any system of short-term forecasting, for these are the only people in a position to make the detailed forecasts required during the construction phase.

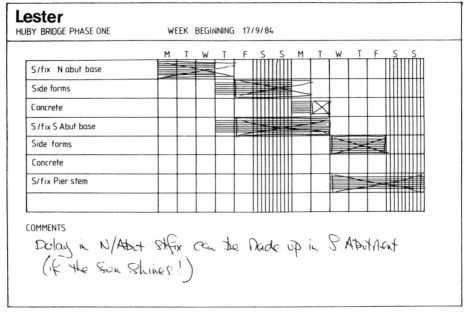

**Figure 17.3** The use of charts for the gathering of short-term forecasting data.

A short-term forecasting system requires the same type of data input as does a reporting system, and is therefore subject to the same needs and difficulties. Again the conflict between accessibility and intelligibility is encountered and again the ideal solution, sophisticated computer equipment in the site cabin, seems to be impractical at present. We are therefore left with the interim solution developed above, a manual system relying on graphics which feeds into a centrally located computer-driven data store.

The manager is now being asked not only what has happened but also what is expected to happen in the next period of, perhaps, a fortnight. Clearly this task is slightly more onerous than straight reporting, but with individually produced forms there is no reason for it to be excessively time consuming. The data are of course self-checking. The forecast is constantly updated as the weekly charts are received, more accurate data are available to the forecaster, and present progress confirms the estimates of previous forecasts. The plan thus takes the form of a changeable forecast in the long term, a highly dynamic forecast in the short term and a static record of the past – another manifestation of the shape of the information wedge which requires more detail in the short term and thus changes as the project ages. Figure 17.3 shows a chart which is suitable for use in a short-term forecasting system.

### The computer-based manual reporting system

The site environment, which is hostile to computer hardware, has required a change of method from one which is purely computer based to one which relies in part on manual methods. As with all manual systems there is a danger that the system will produce paper and little else. The forms must be well designed and produced and must be such that the transfer of information from site to form and from form to computer is quick, accurate, and above all, pleasant.

The information transfer from site to form takes place in the site cabin at the end of the day and this means that the means of transfer must be as quick and as painless as possible. The imaginative use of graphics and the careful selection of the information requirement is important here. The limiting of the required information can effectively be done by the use of the hierarchical structure which has been described earlier in this book, and the limits must be both of scope, i.e. the number of activities, and time.

The information transfer form to computer takes place in the much more controlled environment of the site office and at a much less pressured time of day, the handling of this data is one of the routine daily quasi-clerical tasks to be done. The source of error in such a transfer is not the environment but the personnel, for the work is done by people who have little knowledge of the activities described by the data on the charts and thus will not be in a position to correct errors made by themselves or the original reporting staff. For this reason the computer display must exactly duplicate the form, eliminating the need for any interpretation at this interface. This input requirement implies the possession at site level of a high-quality graphical facility.

One of the advantages of the type of system described here is that it is self-checking – erroneous data emanating from any of the interfaces will produce strange plans for the following week and this error will be noisily brought to the attention of those responsible.

## Providing flexibility

It has been argued elsewhere that planning and control are one, control merely being an extension of planning into the real time of the project. If the manager of the project is to have the flexibility quickly and radically to change the plan it is essential that he should have the appropriate tools. These tools are precisely those required by what has been referred to as the planning phase.

The requirements of the planning phase have been considered elsewhere and need not be repeated; all these requirements remain during the control phase with the additional requirement that the planning tools must accept the new data which are being provided by the project itself. At its simplest this includes the updating of charts and diagrams with the actual times of events and the implications of these actual times, but it may also include automatic updates of future similar production estimates and new resource constraints.

## The future

The mechanism described here is a way of using the new power given by computer hardware to remove some of the difficulties standing in the way of the

## INTRODUCTION OF COMPUTER-AIDED CONTROL

efficient control of construction sites. It deals with that part of the process which, in the past, has given the most difficulty, but does not affect that part, the decision making, which is most dependent on the manager. Clearly it is a very gentle revolution and one which may in the future be seen to have been too cautious. It does, however, attempt to use the new facilities and attempt to avoid the pitfall of the computerisation of inadequate systems.

# 18
# THE NEED FOR AND ACHIEVEMENT OF COST ESTIMATING ACCURACY

Most construction projects in the UK are let by competitive tender and it is in the formulation of these tenders that most planners are first involved with projects. The techniques described in this book have much to contribute to the production of cost estimates, but before considering this it is useful to consider the way in which accuracy is achieved in cost estimation and also the importance of its achievement. For as in all modelling situations, the production of an appropriate model is dependent upon a knowledge of the use it is to be put to.

**Competitive tendering and company policy**

Most contracts in the construction industry are let by competitive tender, and in most cases the successful tenderer will be he who submits the lowest bid. Superficially this seems to be an excellent way to ensure, firstly, that the client gets the cheapest job and, secondly, that experience and enterprise on the part of the contractor are rewarded. This indeed would be the case if all tenders were accurate predictions of the cost of carrying out the project using the methods devised by the contractor, but very rarely is this so.

Tenders are a combination of the technical forecast of the cost of the work to be done and the commercial judgement of the senior management of the company. Thus they can differ from the true cost of doing the work in three ways, due to inaccuracies in the estimating process, to savings inherent in the proposed method, and also to variation in the mark-up applied to the estimate by commercial management. This latter component is difficult to forecast because it depends upon pressures which are internal to the firm. If a company has been very successful in winning contracts it will increase its mark-up to limit future successful bids to those which promise high returns. Conversely a company carrying a small workload seeks, by reducing its margins, to win contracts which merely provide a continuity of activity for its labour and plant. These commercial pressures have the effect of increasing the difficulty of forecasting the level of a competitor's bid.

In addition to the variation in bid which is consciously applied by the tenderer there are variations due to the method of work to be adopted and the accuracy of the estimating.

As stated earlier, the primary justification for competitive tendering is that it gives the client the cost benefits of the experience, expertise, and ingenuity of the tenderer, ensuring that he who best uses his resources of labour and plant, and who is able to negotiate the most favourable rates for materials supply, will be awarded the contract. Unfortunately, in the construction industry there is often very little scope for revolutionary innovation in terms of methods of work, and this advantage to the client may well be limited to those occasions when the contractor is fortunate enough to have plant or labour available in the area at the time required. There is advantage to the client here, but it is not normally decisive in the contract competition.

The third source of variation in the level of bid is the inaccuracy of the estimate. Estimates are by their nature inaccurate, and the tenderer in a typical construction industry competition has a very short time in which to gather the information required. For a large project he may have 12 weeks, but frequently the time is shorter. During this time he must devise a method of carrying out the project together with the outline design of any necessary temporary works, he must familiarise himself with the site, ascertain labour costs and quality, discuss material supplies and collate this information into the figure which is to become his tender. Inaccuracy is inevitable.

Estimates are normally constructed by adding the cost of the various elements of work, and thus there is a tendency to underestimation; the inexperienced tenderer may overlook difficulties and their costs which his more experienced colleague will include in his cost forecast. In a competitive situation the underestimate is awarded the contract, and thus inaccurate estimating may well lead to the award of the contract to the tenderer who has most seriously underestimated the cost of doing the work.

Academic research has, by analysing the distribution of bids, attempted to optimise the mark-up which the tenderer adds to his cost estimate to allow for profit and overheads. Clearly the larger the mark-up the less likely the bid is to be successful, but the larger is the profit should the contract be won. Multiplication of the profits by the probability yields an expected profit which can be used for optimisation.

Although this work can show the effect of mark-up changes, its assumptions of a steady policy on the part of competitors and of estimating accuracy may make it of little practical help to the contractor. Mark-up level is a policy which is constantly changing within organisations depending, as noted previously, on workload, location, and future prospects, and a policy based on an extrapolation of competitors' past actions may not be accurate. Similarly, the mark-up changes envisaged by the models of around ±2% may be completely swamped by the effect of estimating inaccuracy.

## The implications of low bid acceptance

Because low bids win contracts and because estimating is an approximate process, the tenderer can be reasonably certain that, if he is successful in a competitive tender, either he has found a genuine saving which his competitors have not found, or he has underestimated the cost of the job.

On many occasions the contractor will indeed be able to offer a genuine saving; he may have an organisation already in the area, his particular expertise may enable him to organise his work more efficiently than his competitors, or he may be in possession of the particular plant required by the job. If this is the case, then the competitive tender is performing its function, the contractor is rewarded for his foresight and expertise, and the promoter benefits by having his project performed at low cost. In this situation only a major underestimation or a very small or negative mark-up on the part of a competitor would result in the loss of the contract.

When, on the other hand, there is not a clear advantage due to expertise, location, or equipment, and when all contractors intend to use similar methods with the same labour and plant, he who at the tender stage requires least profit (as calculated by the overestimation minus the true cost) is awarded the job. As the smallest bid gets the job, he will in general be successful when the profit (the mark-up plus the accidental overestimation) is small, that is when the accidental overestimation is negative. Therefore the actual profit made by contractors on jobs they actually carry out (the successful bids) is less than the mark-up added to the estimate at the tender stage.

As the estimating accuracy from the companies in a particular competition can be represented by a normal distribution of bids around the true cost, that is an equal chance of under- or overestimation, it can be seen that the chance of a particular bid being successful is increased as the number of competitors is decreased or as the accuracy of the estimates from competitors is increased. In a competition where there are many competitors it is likely that the winning bid will be a considerable underestimate.

In a situation where all competitors have the same mark-up and the same estimating accuracy, a simple analysis can be made which results in a distribution of successful bids.

Figure 18.1 shows the results of such an analysis for various numbers of competitors, and it can be seen, as expected, that there is a decrease in the most likely value of the winning bid as the number of competitors increases. The bid level in Figure 18.1 is shown as the number of standard deviations of the estimate which the bid is below the mean, thus the more inaccurate the general level of estimating within the industry, the larger will be the loss of profit (the lost margin of McCaffer and Harris).

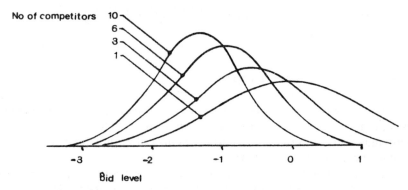

**Figure 18.1** The distribution of successful bids in competitive tenders.

The effect of competition size in reducing the margin of the successful bid is shown directly in Figure 18.2. As noted above the lost margin is indirectly proportional to the level of estimating accuracy.

Obviously Figures 18.1 and 18.2 are based on an extremely simplified model of the tender competition – one in which all companies have the same estimating accuracy, mark-up and true cost. In a real situation these will be different. However, for discussion purposes mark-up can be conflated with inaccuracy and therefore when, as is normal, there is not one company with a significant technical advantage, the lost margins predicted by Figure 18.2 can be seen to be at least a partial reason for the low profitability of the construction industry.

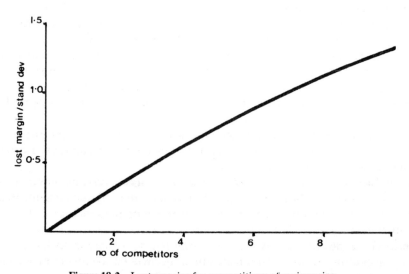

**Figure 18.2** Lost margins for competitions of various size.

## Company policy

It is obvious from the foregoing discussion that the profit margins of the construction industry could be improved if either the general level of estimating accuracy were to be increased or the size of the competition was reduced. Either of these changes would increase the value of the average successful bid. It could be said that such a change would be to the detriment of the promoter, but this is a short-term view; the industry cannot survive with small or negative profit margins and a loss-making contract results either in acrimony and poor quality or in subsequent higher bids.

One company, however, cannot change the whole of the competitive field, and so must act unilaterally; what, it therefore must be asked, is the effect of a unilateral change of company tendering policy?

A company which improves the accuracy of its estimates is less likely to produce the occasional large underestimation which previously had provided its successful bids, and so the change has two effects: it reduces the lost margin of the company on its successful bids and it also reduces the number of bids which are successful. If there were a large number of competitions available and if the cost of tendering were negligible then the reduced success rate would not matter, but this is not the case in the construction industry. A company which wins less than its share of its bids pays for its lack of success in increased overheads. A company would therefore be ill-advised to unilaterally alter its accuracy without also altering the other variable, the mark-up applied to its bids.

A second way that a company can unilaterally change its market share is to change its mark-up. It is, of course, commonplace for a company to do this, it is the usual way in which a company can react to a market share above or below what is desirable. An analysis of the effect of such a unilateral change shows just how effective this is. Just as in the case of a unilateral change in accuracy, a change in mark-up affects both the profit gained and also the success rate, a lower mark-up means a lower profit but a higher success rate and, in a world of high overheads, the higher success rate reduces the pain of the lower profit.

A company has the ability to alter both these tendering parameters and may feel that this would be advantageous. Thus the reduction in market share due to the increase in accuracy can be offset by a reduction in mark-up. Detailed analysis of this option, which is outside the scope of this book, shows that this is indeed a prudent course of action and results in a constant market share with a smaller lost margin. This saving will occur only, however, if the cost of the additional accuracy is not high. Thus greater accuracy must be achieved by more efficient use of existing estimating resources either by concentrating their effort where they are most effective or by introducing new and more efficient estimating techniques.

If estimating can be improved in this way with no or negligible extra cost then, by improving accuracy by a factor of 2, lost margins could be reduced by 30% of the industry estimating standard deviation, or approximately 1% of turnover, a considerable improvement in an industry whose profitability hovers around 5% of turnover.

It cannot be overemphasised that these charts relate to an enormously oversimplified model of the tender completion, that is, that all competitors act in a similar way, and it should be pointed out that the charts as drawn refer only to a six-competitor competition. However, the commercial advantage of improved estimating indicated by the charts is a strong trend reinforced by the fact that they show a penalty if a company unilaterally reduces its estimating accuracy, that is, if a company allow itself to fall behind its competitors in this respect.

Although further work on this model is possible, it would not be of practical use in indicating management policy in detail. No model could usefully handle the complexity of the tender market, but a simple model such as this can indicate the broad policy which is beneficial to the company.

## Achieving accuracy in analytical estimates

No forecast can be without inherent uncertainty; even for the most likely events there is a finite probability that they will not occur. The estimator must manage uncertainty rather than try to eliminate it. Thus the accuracy he strives for is that which is appropriate for the task in hand; he must be aware that accuracy is an expensive commodity and that the extra cost of his estimate must be justified by an extra return.

In the case of estimating for competitive tenders there is, as has been shown, a commercial advantage in estimating more accurately than competitors, and a certain level of estimating accuracy will be justified. For other estimates, for example for estimates which are used as the basis for planning or design decisions, the appropriate level of estimating accuracy may be very low, sufficient only to ensure that the probability level of making the wrong decision is low. The accuracy of estimating in the case of decision making will therefore depend upon the relative value of the courses of action and upon the consequences of taking an erroneous decision.

There is, of course, uncertainty which cannot possibly be resolved. In construction work the external environment is the major source of uncontrollable uncertainty. Although its effects can be reduced, the reduction is not through extra estimating effort, but through extra physical work on site. Thus there is a limit beyond which it is pointless to pursue accuracy, and that limit is the point at which the uncontrollable uncertainty becomes dominant.

In general, estimates are made by analytical means: the event is split into component parts and estimates provided for those, perhaps by further division of the parts into their elements. Thus an estimate is the result of a hierarchy of analysis. Analysing in this way has two beneficial effects: it enables the estimator to consider the detail of the event and thus to eliminate some uncertainty, and also, by dividing the value into subvalues, it automatically reduces the uncertainty of the estimate. This effect, described by Lichtenburg and used by him in his 'successive principle' of estimating, can be shown to increase accuracy in proportion to the square root of the number of components.

The relationship implied here:

$$S = Sx/N$$

assumes that all the components of the estimate are of equal value and that the estimate for each is of the same accuracy. This is clearly not the case where, as is usual, the basis of estimate itemisation is the bill of quantities, for here the value of individual items ranges from single pounds to hundreds of thousands of pounds and clearly the value of the tender is dominated by the value of a small number of very large items.

## Bill itemisation

Civil engineering bills of quantity are usually written to conform to one of the standard methods of measurement. The standard methods take as the basis of the measurement system the quantity of permanent works required by the client. Usually any works of a temporary nature are implied by the permanent work item, and their value is thus included in the permanent works. Thus the formwork required for concrete is included in the item 'supply formed surface'.

The advantage to the engineer in this system is that he can completely avoid a discussion, at the pre-tender stage, with potential contractors about methods and scope of temporary work and can, on receipt of the tenders, compare like with like in tender comparison or scrutiny. For the contractor this method of billing has the disadvantage that it removes the bill, as the basis of his estimate, by one level from his actual plan of doing the work. Shared resources have obviously to contribute to two separate items in the bill, and the estimation of one item depends upon other items which, perhaps, have not yet been considered; an item by item approach is thus not possible.

In spite of this difficulty most estimators use the bill of quantities as the basis for their estimate and, as has been stated, the distribution of value between items plays a considerable part in fixing the accuracy of estimating required.

A bill item total consists of a quantity multiplied by a rate. Each of these will approximate to normally distributed variables and the distribution of the product, the item value, can be expected to conform to a log normal distribution. Plotting the item value distribution for several widely differing bills suggests that this supposition is correct (Fig. 18.3). This type of distribution, of course, implies an unequal distribution of value between items. The well-known and much quoted rule of thumb that 20% of the items of a bill contain 80% of the value has its basis in this distribution of value and would seem, if the hypothesis that bills conform to log normal distributions is correct, to be conservative. In a pure log normal distribution 85% of the value will be contained in 15% of the items!

## COST ESTIMATING ACCURACY

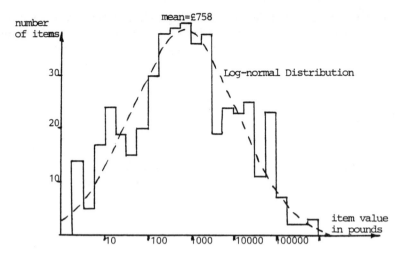

**Figure 18.3** The distribution of item value in a roadworks bill written to the D.Tp method of measurement.

It seems clear that the accuracy of the estimate for individual items in a bill of quantities expressed as a proportion of the value of the items should not be the same for all the items in the bill. The small number of items which dominate the bill should be pursued to a much finer detail than the majority of items which contribute only a small amount to the total. A flexibility of approach is required if valuable estimating resources are to be optimised while simultaneously producing high-quality estimates of cost.

Figure 18.4 shows the effect of this inequality of itemisation on estimate accuracy. A bill is split into 20 items conforming to a log normal distribution.

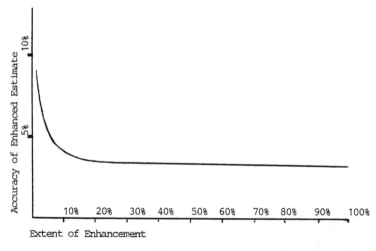

**Figure 18.4** The effect on analytical estimates of partial enhancement.

Two accuracies of estimating are used: an accuracy of ±1 for the small items and of ±0.1 for the larger items. The switchover from accurate to inaccurate is progressively moved down the spectrum of item value. As can be seen the increase in estimating effort is subject to very severe diminishing returns, and an increase in accuracy of the items smaller than those in the largest's 20% has no perceptible effect on the overall accuracy of the estimate.

This simple analysis suggests that estimating accuracy can be best achieved by concentrating on those items which dominate the bill, in this case the 15–20% of the items which have the greatest value. This is clearly a sound strategy but one which, as Lichtenburg has pointed out, can be taken too far, for an individual item can only be as accurate as the uncertainty inherent in the operation, and this may or may not be significant.

## The make-up and accuracy of large value items

The small proportion of large items which dominate a tender consist of either very large quantities of relatively simple parcels of work, e.g. in roadworks, fencing, earthworks, sub-base, or large quantities of parcels of work which are themselves complex, e.g. steelfixing. Both of these classes of work involve multiple repetition of tasks over a considerable period of time and so can themselves be split into discrete parcels of work. This serves to reduce the uncertainty of estimating by averaging good and bad production. An example of this is the effect of weather on earthworks.

It is widely assumed for estimating purposes in this country that one in six days will be lost due to bad weather, so for any given day there is a 16% probability that no work will be done. A binomial distribution can be used to estimate the number of lost days in a given period, and a negative binomial distribution can be used to calculate the mean and the variance of the required construction period:

$$\text{mean} = \frac{\text{required working time}}{\text{probability of work}} = n \times \frac{6}{5}$$

$$\text{variance} = \frac{n(\frac{1}{6})6^2}{5^2} = \frac{6n}{25}$$

$$\text{standard deviation} = \frac{\sqrt{(6n)}}{5}$$

If, therefore, an earthworks activity of expected duration 100 days is being considered, then with great confidence an estimate of its duration can be made of between 105 days and 135 days.

The same analysis can be applied to any repetitious activity subject to random variation, be it the go/no go breakdown of plant, which is exactly analogous to the weather variation examined here, or the random variation in labour productivity which may result in changes of ±50% on daily output but which are reduced by a factor of the square root of the duration when considered for the whole of the activity. Thus it seems likely that the inherent and uncontrollable uncertainties which make accurate estimating very difficult for small-scale work assume less importance for these items which imply numerous repetitions, in general the items which dominate the bill.

In many cases the accuracy of the significant bill items can be increased by buying extra information. It is, however, a matter of some complexity to decide how much information is economic, and at what cost, because usually extra information is costly, and the purchase of information may well be subject to very sharply diminishing returns.

# 19
# COMPUTER-AIDED ESTIMATING

**Cost modelling**

The whole of this book has been concerned in some way with the modelling of projects: first with an examination of the network model and a discussion of its analysis; and then with descriptions of the way in which the newly available hardware can make the handling of the computer model more pleasant and can increase the realism of the models themselves. The means of data capture on site have been discussed, enabling the use of the model to extend beyond the award of the contract into what has previously been called the 'control' phase. In this chapter the adding of cost parameters to the model will be discussed, permitting the production of cost estimates and also providing the cost data which are necessary to test the model during the design cycle.

As has been shown in the previous chapters a model containing project cost data is indispensable at a very early stage in the involvement of the construction team in the project. The UK construction industry relies almost totally on competitive tendering through priced bills of quantity for the appointment of the construction team, and this priced bill of quantity must, therefore, be produced at low cost (because a large proportion of the priced bills will not be accepted and thus the cost of their production will be wasted) and in a short time (because the pricing of bills will always be on the project critical path). Chapter 18 discussed the need for accuracy in cost estimates and the way in which this can be achieved, and has shown (following the work of Lichtenburg) that the detail required of the estimate may not match that provided by the bill of quantities, and indeed that a large proportion of the bill items can be estimated very crudely with little effect on the final estimated sum.

This lack of detail in the early stages of the estimate must, however, be made good at the later stages of the project when it is required, not for the estimation of cost, but for the day-to-day management of the project. The provision of detail in cost estimation, therefore, follows the information wedge which was described in Chapter 1, the need for detail being relatively small until the operations concerned are becoming imminent. The cost model, therefore, is dynamic in the same way that the plan is dynamic, and it is a fallacy that detail is essential in the early stages in order to provide estimating accuracy.

Although the cost model is forced upon the industry by its administration, it is necessary in order to fulfil the purpose which has been emphasised throughout this book, that of design, for the design process implies a test, and cost is the parameter most frequently tested.

In design, as in the provision of tenders, the need for detail in the early stages is unlikely to be great; detail is required merely to enable the planner correctly to make the strategic decisions which will form the skeleton around which the rest of the project plan will be built.

**The achievement of accuracy**

Chapters 12–14 of this book have shown the way in which the increase in detail can be achieved by the use of hierarchical structures. The reason for the introduction of these concepts into the book was primarily to enable the generation and management of managerial decisions, but it is clear that parallel with the generation of details comes the possibility of the generation of precision in cost estimates. In this chapter ways in which cost parameters can be built into the hierarchical model will be described.

Before doing this, however, it is necessary to describe the methods of cost estimating which have traditionally been used in the industry and to examine their implications.

**Methods of estimating**

*Unit rates*

In some respects the bill of quantities which is traditionally used as the basis of tendering and payment in the UK industry is a very useful document. It allows an easy comparison of tenders, not only in terms of their total sum but also in the way in which the sum is distributed, and thus enables the source of discrepancies in tenders to be quickly and accurately identified. The bill of quantities also provides the basis for the payment for work which may vary in quantity (within relatively fine limits) from that which was originally specified, that is a payment system which accommodates the inherent flexibility of the construction project.

In other respects, however, the bill of quantities as at present used is less satisfactory, and in particular this is the case when the bill is used as a basis for cost estimates. The bill of quantities by its nature divides the project into apparently autonomous parts and the tenderer, obliged as he is to price these parts separately in the bill, is encouraged to think of them as entities. If this is done an estimate is built up by a system of unit rates, each rate being the product of the production rate for the item and the cost of the resources employed upon it.

At its best, unit rate estimating can be very sensitive, the estimator building his own experience into both the productivity and the cost figures. At its worst

## METHODS OF ESTIMATING

it becomes an exercise in the uncritical use of past data gleaned either from previous projects or from standard works of reference, neither of which are directly relevant to the project in hand. The strength of unit rate estimating is that it can provide cost estimates at very crude levels of detail. At the preliminary design stage of a project, for example, an estimate of cost may be prepared by producing a 'bill' for the project with very few items, perhaps earthworks, roadworks, bridgeworks and drainage, and this cost estimate can progressively be improved as the work on the project design proceeds, at each stage the overall cost estimate being used to guide decision making. But unit rate estimating contains one flaw which, unless recognised, can seriously reduce its utility; this is the isolation of the component parts of the project.

The division of the project into items in the bill or into separate activities in the network is an artificial one. In almost all cases the resources flow from one sphere of work to another and the people involved neither know nor care which item their work is to be charged to. The economic completion of the project is governed by the efficient use of resources, that is by a smooth flow throughout the project, and this can be achieved only by seeing the project as a whole. Because unit rate estimating ignores the need for smooth flows and thus the cost of uneven flow, it forms a lower bound estimate of cost. The more inexpert the estimator, the larger is the margin by which the estimate falls short of the true cost.

## *Operational estimates*

An alternative to the unit rate estimate is the operational estimate, which ignores the parcels of work as divided in the bill of quantities and looks instead to areas of work which share a particular resource. If operational estimates are to be other than an exercise in unit rates for rather large items, they must be made with reference to a programme of works which fixes the resource use in time. Figure 19.1 shows a histogram for the joiners required for the bridge

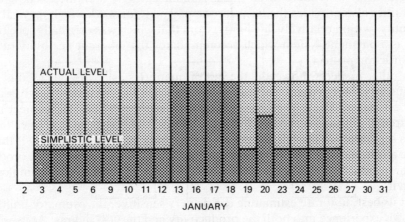

**Figure 19.1** The resource histogram and envelope.

# COMPUTER-AIDED ESTIMATING

construction job referred to in Chapter 8 and, in particular, the way in which the actual provision of joiners differs from that which would be assumed by the unthinking application of unit rate methods.

Operational estimating will provide a more accurate value of total cost than unit rate methods because of the inclusion of at least some of the resource wastage. It does this however at the cost of increased effort, for more detail is required and the successive improvement of accuracy which is a feature of unit rate methods is not possible. Operations estimates tend to form a lower bound because lack of detail implies the possible exclusion of resource use. They present a particular problem for the tenderer – that of mismatch between the elements of the estimate (based on groups of network activities) and the items of the bill. The user of operational estimating methods is forced to allocate his costs between bill items in a semi-arbitrary way, perhaps pro rata with the resource use in the unit rate.

## The role of the computer in cost estimation

In the non-interactive computer environment of the past the computer had little to offer the estimator. A library of standard rates to be applied to the various standard bill items of, for example, the CESMM could be provided and these could then be applied to the bill stored within the computer data storage system. But to call this operation cost estimating is to flatter it, for the essence of cost estimation is the application of the experience of the estimator to the bill, and this is entirely missing from the automatic systems of the past.

In estimating as in all other aspects of CAD the computer must be used to *aid* the designer rather than to force him to use tools which are less than satisfactory. Thus a computer-aided estimating system must incorporate the best currently available methods, and these methods may well be different at different times in the estimating process. An example will illustrate this.

Figure 19.2 shows the hierarchical structure of a highway project, the detail

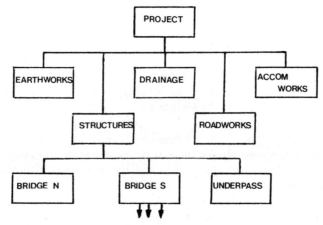

**Figure 19.2**  The hierarchical structure of estimates.

## THE ALLOCATION OF UNIT RATES

being pursued for one of the bridges. For estimating purposes at the tender stage, different degrees of detail are necessary: the large items, first-stage earthworks and roadworks require detailed analysis, a large proportion of the total sum being contained here; the smaller items, the construction of the bridge elements, even of the bridges themselves, can be estimated to sufficient accuracy with only a relatively crude unit rate. The two systems of estimating described above are used side by side and to impose one or the other merely because that is what is contained in the computer package would be a retrograde step.

The need to run alternative methods of estimating side by side together with the insights provided by the hierarchical programs described earlier lead to a system which in principle is easy to use and which, interestingly, leads to an upper bound cost estimate.

## An interactive estimating system

The analysis of the design process which was attempted earlier showed the need, within design, not only to travel around the design cycle but also to descend through a hierarchy of design detail. Lichtenburg has shown that there is a real limit to the utility of this descent due to the inherent uncertainties contained in project plans. If the estimation process is to be part of design (and it must be if cost is to be one of the tested parameters) then a project design package, such as has been described in the early part of this work, must include cost. The discussion above has shown that this inclusion of cost may be in the form of unit rates (applied now not to bill items but to activities or groups of activities within the hierarchy) or of operational estimates. The system must therefore include a method for the inclusion of unit rates, a method of specifying resource use and availability and a method of allocating the costs back to the bill of quantities. These will be described separately.

## The allocation of unit rates

The horizontal histogram, used earlier to provide information for the one specified parameter, duration, is the ideal medium for the provision of cost data. A list of activities is provided as before, now with the opportunity to indicate the activity cost graphically. Alternative approaches are available here: the daily cost of the activity, or the total cost of the operation can be used. The first is most appropriate for the resources of labour and plant, the second for materials, and thus it is helpful to specify these two costs separately on separate histograms.

The use of a daily cost figure rather than a global figure for time-based costs brings two advantages: it enables the separation of productivity (in the duration estimate) and cost (in the daily cost estimate) and thus returns some of the control back to the estimator, and also it allows the use of uncertainty in the activity network to be reflected in the cost estimation.

## The specification of resources

It was shown in the earlier chapter on hierarchies that a knowledge of at least the key resources on the job is necessary if the 'soft' links imposed by management were successfully to be anticipated. Clearly, resources specified in this way for one management purpose can easily be used for another, and the program described here assumes such a double use.

The specification of a resource for an activity implies the commitment to an expenditure equal at least to the product of the daily cost of the resource and the activity duration. This cost will already have been included in the daily cost specified in the histogram described above, and thus the cost of the specified resource must be deducted from the daily cost stored and displayed for the activity. In this way the addition of detail in the form of the specification of resources shifts the method of estimating from the unit rate method to the operational method, but only in those areas where the estimator finds it appropriate. For most activities a combination of both methods is used.

The specification of resources in this way provides, of course, the opportunity of resource allocation and levelling. Automatic systems of resource levelling have been commonplace for many years but have been treated, perhaps justifiably, with caution by planners. An automatic system takes away from the designer the freedom to use his own flair and also, more seriously, creates a model which the designer may not understand, making even the use of a manual override difficult. Experience of the use of the hierarchical programs described earlier suggests that a device such as that described for the specification of soft links makes automatic systems unnecessary, particularly if the display includes both the Gantt chart and the resource histogram.

Using the resource histogram, it is possible both to specify the amount of resources to be made available and also to show the amount of waste. Figure 19.1 showed the resource histogram for a resource together with an envelope diagram representing its strength on site. It is clear that the lightly shaded area beneath the envelope is unproductive time and that this can usefully be represented as a proportion of the time when the resource is in use. The resource envelope is, of course, a design variable and the estimator can, using the graphical means already made familiar by this book, alter the envelope as he wishes, the only constraint being the limits to his action imposed by the need to maintain a stable workforce, the on/off changes for plant, etc. If further specification of resource causes the histogram to go outside the confines of the envelope, then the envelope is returned to its original containing rectangle.

An interesting consequence of this method of resource specification is that it results in an upper bound estimate. The deductions from the unit rates do not include wastage whereas the additions due to operational estimating do and the wastage is greater the cruder is the handling of the resource envelope.

## Pricing the bill

At present there is no alternative for the pricing of the bill to the pro-rata allocation which was advocated for normal operational estimating. The activity costs derived by the method described above include three elements: the cost of materials, the daily rate of those resources not separately specified, and a proportion of the cost of the resources separately specified. This latter cost is of course the product of the daily cost of the resource divided by the productivity and the duration of the activity. All this information is available within the appropriate files.

## Computer-aided estimating

This chapter has described a possible estimating system and must end with a plea similar to that raised in the discussion of control systems. The plea is that the facilities brought by the computer are not squandered in the 'computerisation' of unsatisfactory systems, but that the new facilities are used to permit a radical re-think of methods and an improvement of results.

# APPENDIX
# A GLOSSARY OF GRAPHICS INSTRUCTIONS

The lack at present of a universally accepted set of graphics commands has forced the adoption in this book of a non-standard (although very limited and therefore easy to translate) set of instructions for use in the algorithms used to demonstrate the points raised. The effect of the instructions is:

*block(x1,y1,x2,y2)*     draw a block with horizontal and vertical sides containing the diagonal $(x1,y1)$ to $(x2,y2)$. It is frequently convenient to include in the parameters of such an instruction (which will probably be a procedure of several cruder instructions) a value for the density of the shading or, if available, the colour.

*clear*     clears the screen

*coord(x,y)*     reads the coordinates provided from the screen by the light pen. Such instructions are, of course, hardware specific, depending as they do on the type of facilities provided by the computer.

*eblock(x1,y1,x2,y2)*     erase the block defined

*line(x1,y1,x2,y2)*     draw a line from $(x1,y1)$ to $(x2,y2)$

*move(x,y)*     move the screen cursor to $(x,y)$

*sethor*     set the screen such that the following text is parallel to the $x$ axis

*setvert*     set the screen such that the following text is parallel to the $y$ axis

*write('. . .')*     write the text between the quotes starting at the current cursor position.

# FURTHER READING

Asimow, M. 1962. *Introduction to design*. New York: Prentice-Hall.
Beer, S. 1966. *Decision and control*. New York: Wiley Interscience.
Burmann, P. J. 1972. *Precedence networks*. London: McGraw Hill.
Carr, R. I. 1979. Simulation of construction project duration. *J. Constr. Div. ASCE* **105**, 117–29.
Ehlers, J. H. 1983. The use of colour to help visualise information. *Proceedings CAMP '83*, 158–71.
Jones, J. C. 1976. *Design methods*. London: McGraw Hill.
Koontz, H. and C. O'Donnell 1976. *A systems and contingency analysis of management function*. New York: McGraw Hill.
Lester, A. 1982. *Project planning and control*. London: Butterworth.
Lichtenburg, S. 1974. *Project planning – a third generation approach*. Copenhagen: Polyteknisk Farlang.
Paulsen, B. C. 1976. Concepts of project planning and control. *J. Constr. Div. ASCE* **102**, 587–93.
Snowden, M. 1979. Project management. *Proc. ICE* **66**, 625–35.
Trimble, G. R., H. Neale and S. J. Backus 1979. The effective control of project costs. *Proceedings, Internet '79*.
Twort, A. C. 1972. *Civil engineering supervision and management*. London: Edward Arnold.
Van Slyke, R. M. 1963. Monte Carlo methods and the PERT problem. *Op. Res.* **10**, 839–61.
Wiener, N. 1947. *Cybernetics*. New York: Wiley.

# Index

accuracy of data 85
activity
   at the node 11, *2.2*, 19, 21
   on the arrow 11, *2.2*, 19, 20, *3.4*
allocation sheets 160
allowable stress method 87, 88
analytical estimates 172
Asimow 117

backward pass 23, 51, 54, 55, *6.5*
beta distribution 97, *10.1*, 109, 110, *11.1*
bill of quantities 173, 174, *18.3*, 178
Burmann 128, 130

calendar 40, 41
central limit theorem 97, 102–6
closedown 40, 43, 57, 70
colour 28, 33, 131
competitive tendering 86, 167, 177
complex links 38, 39, *5.1*, 43, 52, 65, 142
complexity 15, *2.4*, 116
computer aided design (CAD) 9, 27, *4.1*, 180
computing cost 8
construction environment 7, 172
continuous processes 5, 6, 152
control 1–4, *1.3*, 5–6, 150–66, *16.2*, *16.3*
cost
   estimates 85, 178
   modelling 177
criteria of success 11, 14, 158
critical
   activities 25
   path 25, 98, 99, *10.2*
   path method (CPM) 10, 19, 20, 22, 62
crosslinks 129, 144, 147
cybernetics 1, 150

data
   input 26, 33, 62–70, 111
   output 113–14
   storage 108, 133–9, *14.8*
decision nodes 89, 90–3, *9.3*
dependency matrix 35, *4.7*, 36, 64–7, *7.5*, *7.6*, *7.7*, *7.8*, *7.9*, 78, 142, 148
design
   cycle 9, *2.1*, 15, *2.3*, 105
   data 5, *1.4*
   detail 16, 117
   process 3, 4, *1.3*, 9–16, 117, 118
deterministic data 86, 88, 93, *9.4*

divisional organisation 125, 126, *13.1*
dummy arrows 20, 21, *3.5*
duration
   distribution 89–91, *9.2*
   histogram 34, *4.6*, 63–4, *7.4*, 76

events 17

feedback 3, 6, 151, 155
feedforward 3, 6, 151
finish date 50
float 25, 88
forward pass 23, 24, *3.7*, 49–53, *6.3*, *6.4*, *6.5*, *6.6*, 144
functional organisation 126, *13.1*

GERT 89, 92
graphical
   input 33–6, 76–81, 111–13, 130, 161
   output 28–33, 62, 81–4, 113–14, 161
graphics 8, 26–37, 43, 74–84
   language 74

hard links 130
hierarchy 14, 15, *2.3*, 39, 59, 92

incentive bonus 159
index 138–9
information wedge 2, *1.2*, 16, 85, 121, 177
input error 33
interactivity 13, 74

Jones 117
journey analogy 1, *1.1*, 150

Koontz and O'Donnell 3, 151

lags 39, 52, 81
leads 39, 52, 81
Lichtenburg 87, 117, 120, 173
lightpen 76
Likert 122
limit state design 12, 88
line printers 28, 29, *4.2*
linkage matrix 131
linked
   lists 136, 137, *14.5*, *14.6*
   rings 136
linkpin 122
logic links 130, 138
lost margin 169, 170, *18.2*

# INDEX

McCaffer and Harris 169
management span 121, 122, *12.3*
mark up 167
mathematical models 10
matrix organisation 127, *13.2*
menus 36, 59, 60, *7.1*, 61, *7.2*, *7.3*, 74, 75, *8.1*, 141, *15.1*, 142, *15.2*
modelling 10–11, 17–19, 62, 177
modify 10, 13
Monte Carlo simulation 96, 99
monthly measure 160

network analysis 5, 10, 17–25
number of simulation cycles 104, *10.7*, 105

operational estimates 179

PASCAL 45, 140
plotter 28–30, *4.3*
precedence diagram method (PDM) 10, 19, 21, *3.6*, 22, 38, 62, 63
 notation 23, 24, *3.7*
productivity 89
program
 direction 59
 evaluation and review technique (PERT) 10, 19, 20, 22, 91, 96–101
project calendar 41

realism 38, 44
resource
 allocation 182
 flow 128
 histogram 179, *19.1*, 182
 links 130, 138

resources 88, 128, 142, 179
response time 3, 7, 150, 153, 154, 161

seasonal variations, 42, *5.2*, 43, 57, 70
short-term forecasting 6, *1.5*, 152, 163
simulation techniques 101, 104, *10.7*, 105, *10.8*, 107
soft links 130, 143
sort 45, 46, 47, *6.1*
steering 1
successive principle 173

tender comparison 173
test 10, 11–12
time saving 27
tree structures 118
triangular distribution 110, *11.1*
Trimble 152
Twort 159

uncertainty 7, 85–95, 96–106, 107–15, *11.3*, *11.5*, 172
unit rates 178, 181

Van Slyke 105

weather 42, 87
Wiener 1
window times 41, 43, 51, 70, 71, *7.9*

zones of equal probability 101, *10.5*, 102